THE BEDFORD SERIES IN HISTORY AND CULTURE

Charles Darwin and the Question of Evolution

A Brief History with Documents

Related Titles in

THE BEDFORD SERIES IN HISTORY AND CULTURE

The Scientific Revolution: A Brief History with Documents
Margaret C. Jacob, *University of California, Los Angeles*

UTOPIA by Sir Thomas More
Edited with an Introduction by David Harris Sacks, *Reed College*

The Enlightenment: A Brief History with Documents
Margaret C. Jacob, *University of California, Los Angeles*

DISCOURSE ON THE ORIGIN AND FOUNDATIONS OF INEQUALITY AMONG MEN by Jean-Jacques Rousseau with Related Documents
Edited with an Introduction by Helena Rosenblatt, *The Graduate Center, City University of New York*

CANDIDE by Voltaire
Translated, Edited, and with an Introduction by Daniel Gordon, *University of Massachusetts Amherst*

NOTES ON THE STATE OF VIRGINIA by Thomas Jefferson with Related Documents
Edited with an Introduction by David Waldstreicher, *Temple University*

ON LIBERTY by John Stuart Mill with Related Documents
Edited with an Introduction by Alan S. Kahan

THE BEDFORD SERIES IN HISTORY AND CULTURE

Charles Darwin and the Question of Evolution

A Brief History with Documents

Sandra Herbert

University of Maryland Baltimore County

BEDFORD / ST. MARTIN'S Boston ◆ New York

For Joann, Bob, Kristen, John, Sonja, and Joel

For Bedford/St. Martin's

Publisher for History: Mary V. Dougherty
Director of Development for History: Jane Knetzger
Senior Editor: Heidi L. Hood
Developmental Editor: Ann Kirby-Payne
Editorial Assistant: Jennifer Jovin
Production Supervisor: Andrew Ensor
Executive Marketing Manager: Jenna Bookin Barry
Project Management: Books By Design, Inc.
Index: Books By Design, Inc.
Text Design: Claire Seng-Niemoeller
Cover Design: Andrea M. Corbin
Cover Art: *Charles Darwin*, 1840, by George Richmond. J980057. © English Heritage Photo Library. By kind permission of Darwin Heirlooms Trust.
Composition: Achorn International
Printing and Binding: Haddon Craftsmen, Inc., an RR Donnelley & Sons Company

President: Joan E. Feinberg
Editorial Director: Denise B. Wydra
Editor in Chief: Karen S. Henry
Director of Marketing: Karen R. Soeltz
Director of Production: Susan W. Brown
Associate Director, Editorial Production: Elise S. Kaiser
Manager, Publishing Services: Andrea Cava

Library of Congress Control Number: 2010932772

Manufactured in the United States of America.

6 5 4 3 2 1
f e d c b a

For information, write: Bedford/St. Martin's, 75 Arlington Street, Boston, MA 02116 (617-399-4000)

ISBN-13: 978-0-312- 47517-8

Acknowledgments

Acknowledgments and copyrights are continued at the back of the book on page 132, which constitutes an extension of the copyright page.

Distributed outside North America by PALGRAVE MACMILLAN.

Foreword

The Bedford Series in History and Culture is designed so that readers can study the past as historians do.

The historian's first task is finding the evidence. Documents, letters, memoirs, interviews, pictures, movies, novels, or poems can provide facts and clues. Then the historian questions and compares the sources. There is more to do than in a courtroom, for hearsay evidence is welcome, and the historian is usually looking for answers beyond act and motive. Different views of an event may be as important as a single verdict. How a story is told may yield as much information as what it says.

Along the way the historian seeks help from other historians and perhaps from specialists in other disciplines. Finally, it is time to write, to decide on an interpretation and how to arrange the evidence for readers.

Each book in this series contains an important historical document or group of documents, each document a witness from the past and open to interpretation in different ways. The documents are combined with some element of historical narrative—an introduction or a biographical essay, for example—that provides students with an analysis of the primary source material and important background information about the world in which it was produced.

Each book in the series focuses on a specific topic within a specific historical period. Each provides a basis for lively thought and discussion about several aspects of the topic and the historian's role. Each is short enough (and inexpensive enough) to be a reasonable one-week assignment in a college course. Whether as classroom or personal reading, each book in the series provides firsthand experience of the challenge—and fun—of discovering, recreating, and interpreting the past.

Lynn Hunt
David W. Blight
Bonnie G. Smith
Natalie Zemon Davis
Ernest R. May

Preface

Charles Darwin is well known, and his name is associated in the public mind with the idea of evolution. What is less well known is the background for his decision to embrace the idea of evolution. This book makes the history of the idea of evolution, from its inception to its acceptance as a scientific theory, accessible to undergraduate students. The story begins in the 1780s, when the question of evolution began to gain shape and urgency, and it ends with Charles Darwin's death in 1882, by which time the idea of evolution had become an established theory in science. But the primary focus of this volume is on the events leading up to and immediately following the 1859 publication of Darwin's seminal work, *On the Origin of Species.*

This book takes a broad view of the history of evolutionary theory and reveals how it is intertwined with other aspects of human history and culture. Taken together, the introductory essay and the documents in this volume help students understand how evolution as a theory depended on the exploration of new lands—especially the New World—which in turn depended on political expansion. Many of the figures discussed in this book were travelers who were in one way or another supported by their home governments. Further, the collecting of exotic plants and animals, living or fossilized, required the creation of museums to house the new specimens. All of these developments took place in a global context involving the relationship of peoples of many lands. Hence the subject of evolution became entangled with such political issues as the contemporary campaign to abolish slavery. This book allows students to unravel the skein of these entanglements.

This book is intended for use in university classrooms in courses that deal with history and science. Part One's thorough introduction provides students with the context needed to understand the world in which Charles Darwin launched his theory. It outlines the intellectual and personal connections Darwin had with the theorists who came before him. Part Two provides primary sources for students to engage on their own. The twenty-nine documents contained in this book display the wide variety of materials on which historians customarily draw in interpreting the past. They include intimate letters, sketches and illustrations, and photographs of objects as well as formal scientific texts. The first group of documents (Chapter 1) represents the period from the 1780s through the 1830s, when the idea of evolution in its modern form was first broached. The second group of documents (Chapter 2) focuses on the development of the idea of evolution among Darwin and several of his important contemporaries. Each document is introduced with a thoughtful headnote that helps students understand the item's overall importance as well as its specific connections to Darwin and his theory. Footnotes clarify references within documents that may be unfamiliar to students.

Additional pedagogical aids support students' synthesis of the materials in this volume. A useful chronology charts the development of the theory of evolution over several decades. Questions for consideration ask students to think critically about the documents. Finally, the selected bibliography contains an abundance of sources for further student research.

ACKNOWLEDGMENTS

I would like to express my gratitude to Lynn Hunt for inviting me to contribute to the Bedford Series in History and Culture. At Bedford/St. Martin's, I would like to thank Mary Dougherty and Heidi Hood for fruitful conversations and Jennifer Jovin for editorial support. I am grateful to Ann Kirby-Payne for a careful reading of my text, and I offer my thanks to Emily Berleth at Bedford/St. Martin's and Nancy Benjamin at Books By Design for seeing the manuscript through the production process and to Judith Riotto for a fine copyedit. I am also grateful to the following instructors, who reviewed the manuscript and provided useful feedback: Paul Farber, Oregon State University; Robert Friedel, University of Maryland, College Park; Katherine Pandora, University of Oklahoma; Michael Ruse, Florida State University; Shirley Roe, University of Connecticut; and James Sack, University of Illinois at Chicago. As always I am deeply grateful to my husband, Jim Herbert, for good conversation and counsel. I say a warm thank-you to those at Christ's College at the University of Cambridge, where I spent the academic year 2006–2007 as a visiting scholar thinking about this book, and to my students at University of Maryland Baltimore County, who participated in a seminar in the fall of 2007 trying out a scheme for the book. To my longtime colleagues in the Department of History and in the Program in the Human Context of Science and Technology at UMBC, I extend my good wishes and gratitude. I am also grateful to my teachers at Wittenberg University and at Brandeis University for introducing me to the fascinating subject of the history of evolutionary ideas.

As has been true throughout my career, I am indebted to the magnificent collections of the Library of Congress and to its excellent staff, and most especially to Constance Carter, head of the Science Reference Section. At Cambridge University Library I am particularly grateful to Adam Perkins of the manuscripts department. To the staff of the Darwin Correspondence Project, and particularly to Alison Pearn, I say thanks. Peter J. Gautrey and David Kohn, my colleagues in editing Darwin's notebooks, have been friends for many years, and I thank them. Thanks also to Alan Leviton of the California Academy of Sciences for photographing the plates of Darwin's mockingbirds.

Sandra Herbert

Contents

Map and Illustrations

MAP

ILLUSTRATIONS

THE BEDFORD SERIES IN HISTORY AND CULTURE

Charles Darwin and the Question of Evolution

A Brief History with Documents

PART ONE

Introduction: Development of the Theory of Evolution

Charles Darwin (1809–1882) is the central actor in our story. Darwin's book *On the Origin of Species*, published in 1859, inaugurated a new era in the biological sciences. By 1900 the theory of evolution had been accepted, with variations in emphasis, by a majority of scientists worldwide. In his book Darwin explained how new species might come into existence. His theory held that new species were the descendants of previously existing species. Thus his theory was one of descent with modification. The process by which species changed he called *natural selection*. Although Darwin was an innovator, his ideas rested on a foundation of fact and analysis that naturalists and explorers around the world had built up over the course of a century or more.

SOURCES OF EVOLUTIONARY IDEAS IN THE LATE EIGHTEENTH CENTURY

From the European side of the story, work done outside the landmass of the Old World was critical. Many of the key discoveries that led naturalists to posit a theory of evolution were made in the Americas, in Australia, and on islands in the Pacific Ocean. Although the great age of European exploration began in the fifteenth century, it was not until the eighteenth century that discoveries in natural history were fully exploited. One person whose work facilitated the understanding of species was Swedish naturalist Carl Linnaeus (1707–1778), who set up the system of binomial nomenclature whereby every species in the world was to be assigned a genus and species name in Latin. For example, the Linnaean name for human beings is *Homo sapiens*.[1] This system of naming remains in use today. It provides naturalists with a common language to use when discussing and comparing species.

In addition to setting up a universally applicable system for naming species, Linnaeus set species within a hierarchy of taxonomic categories: Genera were grouped into orders, orders into classes, classes into kingdoms. Linnaeus was also interested in collecting new species. Because the greatest supply of new species was from outside Europe, Linnaeus made it a point to train students in the techniques of classification and then arrange for them to travel abroad before returning to Europe with novel specimens. One of his most successful students was Daniel Solander (1733–1782), a Swede who traveled with Captain James Cook (1728–1779) in a voyage to the Pacific (1768–1771). In effect, Linneaus promoted internationalism in science on two levels: in his use of Latin for naming species and in the placement of his students. An English translation of one of Linnaeus's works was done anonymously by Charles Darwin's grandfather, physician and poet Erasmus Darwin (1731–1802). In a gesture of transatlantic friendship, Erasmus Darwin sent a copy of the book to American statesman and inventor Benjamin Franklin (1706–1790) (Document 1).

Another great eighteenth-century figure in natural history was French naturalist Georges-Louis Leclerc, Comte de Buffon (1707–1788). Buffon's energies and his corresponding contributions to the field were remarkable. He supervised the writing of an enormous thirty-six-volume exposition of natural historical knowledge entitled *Histoire naturelle, générale et particulière*, which was published between 1749 and 1789. In outlook Buffon presumed a very long temporal dimension for the earth, a subject that was only beginning to become important in European thought. Relying only on ancient texts, including the Christian Bible and what was known of ancient Chinese and Egyptian writings, most educated men and women during Buffon's time imagined an earth that was about six thousand years old. While it would not be until the twentieth century that the age of the earth could be measured accurately, even a sense that the earth was very old represented a change in perspective. Buffon also directed the most important institution in Europe devoted to natural history, the Jardin du Roi (the King's Garden) in Paris (renamed the Muséum National d'Histoire Naturelle during the French Revolution). Buffon's accomplishments were widely heralded in the salons of eighteenth-century Paris, and natural history assumed an honored place in public life.

By 1790 Linnaeus and Buffon had died, but their work was continued by other naturalists throughout Europe. At the turn of the nineteenth century, from 1799 to 1804, Alexander von Humboldt (1769–1859), a native of Berlin, explored a large swath of the northern and western portions of South America for the purpose of making observations regarding its geography and natural history. He was inspired to travel by reading accounts of Cook's travels. In turn, he encouraged others to explore. For example, in 1804, as the Lewis and Clark expedition was being prepared, Humboldt counseled President Thomas Jefferson (1743–1826) on aspects of that mission.[2] In a similar vein, some thirty years later, Humboldt's account of his travels inspired the young Charles Darwin to travel for scientific purposes. As an undergraduate, Darwin read Humboldt's account of his travels to the Southern Hemisphere and was moved by passages describing the beauty of distant places (Document 2).

The late eighteenth century also yielded another crucial development in the realm of plant and animal breeding. The goal of this work was largely practical—that is, to increase agricultural yields or, as in the breeding of thoroughbred horses, to increase speed. However, the success of domestic breeding programs demonstrated that new breeds of animals and new varieties of plants could be produced artificially. Once the subject of mutability in species had been broached in the late eighteenth century (stimulated largely by speculation over species extinction), those thinking about evolution could draw on the fund of knowledge regarding domestic breeding.

Progress in agriculture was matched by progress, or the hope of progress, in other areas of life. The invention of the steam engine laid the foundation for the industrial revolution. A sense of the possibility of improvement extended to the realm of politics. Was monarchy the only legitimate form of government? Was republican government—in which power rests ultimately with the governed—a possible alternative? In America, Thomas Jefferson and his colleagues had debated these points in the 1770s and 1780s, and they looked for guidance from the "laws of nature." The intertwining of interest in nature and in society is best captured in this collection by an excerpt from a work by Erasmus Darwin. Its title indicates its broad reach: *The Temple of Nature; or, the Origin of Society* (Document 3).

Men like Jefferson and Erasmus Darwin favored republican government. More generally, they believed in the notion of the perfectibility of human society, even while remaining aware of human imperfections, including their own.[3] One point on which they differed, however, was regarding the abolition of the institution of slavery. While Jefferson disliked the institution of slavery, he did not call for its immediate abolition. In contrast, Erasmus Darwin was a keen abolitionist, as was Charles Darwin's maternal grandfather, Josiah Wedgwood (1730–1795). Indeed, Wedgwood, who founded the Wedgwood pottery works, contributed an often-reproduced ceramic medallion to support the abolitionist cause (Document 4).

Thomas Jefferson's pursuit of learning is well known. One of Jefferson's teachers at the College of William and Mary in Virginia was Scottish-born doctor William Small (1734–1775). Small eventually returned to Britain and played an important role in forming the Lunar Society, a group of intellectuals that included Erasmus Darwin and Josiah Wedgwood. Thus Jefferson, Erasmus Darwin, and Josiah Wedgwood were linked by the tie of common associations.[4] Jefferson's involvement with the culture of the French Enlightenment was even greater, for he resided in France from 1784 to 1789 while serving in official capacities representing the American government. While he was in France he published his only book—*Notes on the State of Virginia* (London edition, 1787)—as a response to a series of encyclopedic questions posed by a French diplomat.[5] Among the subjects he discussed in the book was the natural history of the United States. Citing numerous examples, Jefferson refuted claims by such European naturalists as Buffon that American species were inferior in size to European species. He also called particular attention to the existence of fossil bones of a gigantic elephant-like tusked quadruped. He remarked that the animal could not have become extinct for that would be in contradiction to the economy of nature (Document 5).

Continuing his work on fossil animals, in 1797 Jefferson read a paper to a meeting of the American Philosophical Society in Philadelphia describing a new animal from rural Virginia that he called the "Great-Claw" or *Megalonyx*.[6] Following Indian tradition, Jefferson suggested that the animal was still living elsewhere on the continent.[7] Jefferson hoped that the Lewis and Clark expedition to the West (1804–1806) would find living representatives of both the elephant-like animal and the "Great-Claw." No such animals were found. Eventually the first groups of bones described in his *Notes on the State of Virginia* of 1787 would be identified as two different extinct species: *Mastodon americanus* and *Elephas primigenius*. The gigantic sloth-like "Great-Claw" he described in 1799 would be named after him: *Megalonyx jeffersonii*.[8] In interpreting fossil bones, Jefferson deferred to the authority of French comparative anatomist Georges Cuvier (1769–1832), a curator at the Muséum d'Histoire Naturelle.

When he did his work on fossil bones, Jefferson was writing on the cusp of change. In the 1790s Cuvier was in the process of establishing the reality of species extinction with his study of fossil quadrupeds (Document 6). Cuvier's impact on Jefferson appears in a postscript, dated March 10, 1797, that Jefferson attached to his publication on the "Great-Claw." Jefferson referred to a new find by Cuvier, called the *Megatherium americanum*, a notice of which had been translated from the French and published in London's *Monthly Magazine* in September 1796. Describing the discovery, Cuvier wrote:

> It adds to the numerous facts which apprize us that the animals of the ancient world were all different from those which we now see on the earth; for it is scarcely probable, that if this animal still existed, so remarkable a species could have hitherto escaped the researches of naturalists.[9]

Jefferson did not immediately embrace Cuvier's new views on extinction, and thus his own work was quickly outdated and surpassed. Yet Jefferson's contributions to vertebrate paleontology in the United States were important. He set an example for others by his very public attention to the subject: He used what is now the East Room of the White House to store fossils. Jefferson also exchanged fossils with others, in his own country as well as in France.

Erasmus Darwin figures prominently in the progression shaping thought about the natural world in the late 1790s and early 1800s. He sought to enlist "Imagination under the banner of Science; and to lead her votaries from the loose analogies which dress out the imagery of poetry, to the stricter ones which form the ratiocination of philosophy."[10]

In other words, he envisioned his work, which mixed poetry and prose, as a bridge between the "looser analogies" of poetry and the more reasoned discourse of philosophy. This mode of presentation was too idiosyncratic to become a standard style in the sciences, but it did enlarge the imagination of his readers and lay the foundation within the family for another evolutionist—his grandson Charles Darwin.

On the scientific side, Erasmus Darwin's interests spanned the three traditional branches of natural history: geology, zoology, and botany. His insights were derived from extensive reading, from his own education and practice as a physician, and from participation in the meetings of the Lunar Society. He was intrigued by the fossils his colleague Josiah Wedgwood sent him from canals then under construction, and he was stimulated by writings of their elder colleague John Whitehurst (1713–1788). Darwin also drew on the writings of James Hutton (1726–1797) and Buffon, both of whom emphasized the vastness of geological time. Darwin presumed that some species had become extinct and that others had replaced them, though he offered no proof. In zoology his own training as a physician was paramount, and he drew on recent research in chemistry to enhance his views on physiology. In botany he made his own most concrete contribution, for he brought Linnaeus's classifications of plants to an English-speaking audience.[11] Something of the spirit of Erasmus Darwin's thought (as well as his fondness for American geography) can be gathered from the following text:

> For Vegetables are, in truth, an inferior order of Animals, connected to the lower tribes of Insects, by many marine productions, whose faculties of motion and sensation are scarcely superior to those of the petals of many flowers. . . . Thus is the great expanse of Organized Nature divided into districts, and distinguished by names; but, as it branches over her mighty continents, like the lakes Ontario, Erie, Huron, and Superior, each flows into the other by some narrow communication, forming one whole with the wide ocean of created Being.[12]

The connectedness of all forms of life—an idea critical to evolution as a scientific theory—was Erasmus Darwin's theme. As the sum of his writings show, he believed the connection was developmental. One citation from his final work, *The Temple of Nature*, will suffice to make this point:

> The mystery of reproduction, which alone distinguishes organic life from mechanic or chemic action, is yet wrapt in darkness. . . . But it may appear too bold in the present state of our knowledge on this

> subject, to suppose that all vegetables and animals now existing were originally derived from the smallest microscopic ones, formed by spontaneous vitality? And that they have by innumerable reproductions, during innumerable centuries of time, gradually acquired the size, strength, and excellence of form and faculties, which they now possess? And that such amazing powers were originally impressed on matter and spirit by the great Parent of Parents! Cause of Causes! Ens Entium![13]

Although the word *evolution* had not yet been applied to this notion, the direction of Erasmus Darwin's line of thought is clear. His grandson Charles Darwin would carry this line of thought further.

EVOLUTIONARY AND ANTIEVOLUTIONARY CURRENTS IN EARLY NINETEENTH-CENTURY THOUGHT

The next stage in our story is a complicated one. Intellectual opinion in Europe was affected by the French Revolution and by the subsequent rise to power of Napoleon Bonaparte (1769–1821). Reaction against the ideals of the Revolution was strong in Great Britain. As we have seen in the writings of Erasmus Darwin, notions of progressive change in the species world were linked with notions of progressive change in the political world. Some observers claimed that such ideas contributed to the revolutionary turmoil in France. There was thus a turning away from ideas such as those Erasmus Darwin espoused. In 1795 he was aware of this disapproval and considered following another Lunar Society member—chemist Joseph Priestley (1733–1804)—to the New World. Gripped by antimonarchical and anticlerical (though not necessarily antireligious) sentiment, Erasmus Darwin wrote to a friend, "America is the only place of safety—and what does a man past 50 want. . . . Potatoes and milk—nothing else. These may be had in America, untax'd by Kings and Priests."[14]

The conservative reaction to the French Revolution attracted some who were repelled by the violence they saw happening across the English Channel and began to rethink its claims about the perfectibility of human society. For example, Thomas Robert Malthus (1766–1834) turned away from the optimistic views of his parents' generation to embrace a more pessimistic view of human nature. Like Erasmus Darwin, Malthus was a graduate of the University of Cambridge, and the two men even shared a publisher—J. Johnson in St. Paul's Churchyard, London.

However, where Erasmus Darwin was sympathetic to the goals of the French Revolution and to the French Enlightenment, the more conservative young Malthus was skeptical. He announced his views in *An Essay on the Principle of Population* (1798) (Document 7). There he noted "that tremendous phenomenon in the political horizon, the French revolution, which, like a blazing comet, seems destined either to inspire with fresh life and vigor, or to scorch up and destroy the shrinking inhabitants of the earth."[15]

In the course of offering a critique of the thinking of several recent authors, Malthus identified what he believed was the limit to human perfectibility: the tendency of human populations to rise exponentially unless checked by such constraints as late marriage. Malthus's conservative views became fundamental in the new social science of economics. His work also helped to create a conservative intellectual mood in Britain during the first three decades of the nineteenth century. Ironically, years later, Charles Darwin would redirect the Malthusian understanding of population dynamics to explain how species adapt.

Traditional religious views revived in the aftermath of the French Revolution. For example, William Paley (1743–1805) argued in his *Natural Theology* (1802) that nature's "contrivances" (that is, adaptations) were arguments in favor of the existence of God.[16] Like Erasmus Darwin, Paley was a graduate of the University of Cambridge, and he was well aware of his fellow Cambridge author. He referred to Erasmus Darwin's views of a self-generating nature in a polite but skeptical tone.[17] Paley viewed evidence of design in nature as argument for the prior involvement of a designer. He made his point in the opening paragraph of *Natural Theology* (Document 8). Paley paid such careful attention to the elaborate adaptations observable in nature that when Charles Darwin came to work on the subject of evolution, he knew that, above all, he had to offer an explanation for them.

Alongside the tumult in politics, during the early nineteenth century there was significant progress in various branches of natural history, including geology. The Geological Society of London was formed in 1807 and, being private rather than governmental in origin, was a characteristically British institution. Within twenty-five years of the society's founding, British geologists were working out the fundamental sequence of British strata.[18] In the early decades of the nineteenth century, the educated public gradually became used to thinking of life on earth as having been of long duration, though no absolute numbers were yet applied. Still the days were over when even such an educated man as Malthus, writing in 1798, could refer to the "five or six thousand years that the world has existed."[19] From 1790 to 1815 British naturalists also continued to add to public and private collections, those at the British Museum being important but by no means the only significant collections. Certain groups of specimens such as hitherto unknown marsupials from Australia, were of great contemporary interest, and soon would play a prominent role in discussions of the geographic distribution of species. It was also helpful to naturalists that botanist Joseph Banks (1743–1820) was president of the Royal Society of London from 1778 to his death. He was instrumental in supporting the institutional development of natural history (as in making Kew Gardens a truly scientific endeavor) and in furthering the cosmopolitan nature of scientific pursuits. He also managed the exquisitely difficult task of maintaining ties among scientists of warring nations during the Revolutionary and Napoleonic periods. Still, when all was said and done, it was in France, rather than in Britain, where the subject of evolution found its first real home.

The commitment of France to good science endured through tumultuous political upheaval. How was such stability of purpose possible? The answer rests in the strength of the primary French institution devoted to natural history: the Muséum d'Histoire Naturelle. This institution operated in the tradition of French science. Unlike British institutions devoted to science, French institutions had significant government funding. A key ingredient in assuring stability was the museum's policy regarding appointments. After 1793, there were twelve permanent professorial positions. Because professors held tenure and salary for life, they could hold and defend strong opinions. This proved to be important for the advancement of natural history.

Furthermore, the institution combined numerous functions. As historian Dorinda Outram has pointed out, "In the later eighteenth century the Muséum had been largely a botanical garden; after 1793 it began to be an institution which remained an outdoor setting for the display of nature but also began to house increasing numbers of built spaces like galleries and dissection rooms to house an increasingly indoor science."[20] By "indoor science" Outram is referring to what we would today call reference collections, that is, collections so vast that their cataloging and classification could be authoritative for general theoretical conclusions for the entirety of the natural world. It was increasingly seen as the responsibility of the museum's professors and their staff to organize these collections. There was considerable jockeying for space to house them, a process made the more complicated by the fact that staff lived on the grounds.[21] Whatever the difficulties of storage, it was the combination of highly qualified professional staff working with superb reference collections that allowed the naturalists of the museum to come to the judgments about species that they did.

Of the professional naturalists at the museum, we will give primary attention to two: comparative anatomist Cuvier (already discussed) and his older colleague Jean-Baptiste Lamarck (1744–1829). They are often presented as polar opposites, and in some ways they were. While Cuvier was a master diplomat who navigated the corridors of power with great skill, Lamarck was, relatively speaking, politically inept. While Cuvier advanced a view of classification that separated major groups of animals into distinct branches, Lamarck adhered to an older image of a "great chain of being" that placed animals in a series, albeit now with an evolutionary connection threading together the chain. Lamarck's reworking of the old notion of a "great chain of being" reinforced the notion of common origin (Document 9). While Cuvier is known as an ardent opponent of evolutionary thinking, Lamarck is credited for his advocacy of the idea. Still, as colleagues at the same institution, Cuvier and Lamarck were united in their deep attachments to the collections of specimens under their control.

From his studies of fossil bones and his deep knowledge of vertebrates worldwide (made possible only by the extent of the museum's collections assembled from all continents of the earth), Cuvier deduced that some of these animals had become extinct. Further, as the result of his deep knowledge of comparative anatomy, Cuvier was able to suggest which animals were most like the extinct animal. A critical specimen in his argument was the *Megatherium* collected in an area to the west of the city of Buenos Aires in present-day Argentina. This fossil skeleton was unusual in that it was nearly complete. The mounted specimen was in Madrid, where drawings were made of it. Cuvier was called upon to interpret the anatomy of the skeleton. He described it as different from all known living animals but with a family-like resemblance to currently living sloths, a group present in South America (Document 6).

The essential geological element to Cuvier's interpretation was that he assigned the *Megatherium* to a prehuman world. Further, he rejected any kind of evolutionary change to account for the relationship between

Megatherium and living sloths. He believed that the magnitude of the difference between extinct animals and their nearest living relatives was too great to be explained by climatic change. His conclusion was that such animals as *Megatherium* had become extinct through some kind of geological catastrophe. In the next twenty years, specimens from other extinctions were rapidly unearthed, including many from the region around Paris itself. Cuvier collected his individual researches into a four-volume work published in 1812. Cuvier's masterwork was entitled *Recherches sur les ossemens fossiles de quadrupèdes* (*Researches on the Fossil Bones of Quadrupeds*).[22]

The topic of species extinction was equally important to Lamarck, though he took it in a different direction. Aware of his colleague Cuvier's researches on the fossil bones of quadrupeds, he faced similar problems in the study of invertebrate animals—particularly the study of shells (conchology), which was his specialty. Had some species of fossil shells become extinct or not? As historian Richard Burkhardt has put it,

> By 1799 it was Lamarck who was looked upon as the man with the expertise in conchology. I believe that it was the specter of species extinction, as posed by the differences between fossil and living forms, that drove Lamarck to his belief in species mutability in the first place.[23]

As Burkhardt pointed out, quoting an 1801 paper by Cuvier, there were basically three ways of explaining the differences between fossil and living forms: the destruction of the form, a modification of the form, or migration of the fossil form, before its destruction, to a region of the globe with a different climate. Cuvier relied on the first and third of these explanations; Lamarck on the second and the third. Thus, the question of the mutability of species first arose, in its modern form, in the context of extinction and migration.

Lamarck first articulated his views in a lecture in 1800, but his most extended treatment of the subject was in a book entitled *Philosophie zoologique* (*Zoological Philosophy*), published in 1809.[24] On the basis of his book Lamarck is credited, and properly so, with being the founder of evolutionary theory. However, it should be pointed out that he did not use the term *evolution* or indeed any single word or phrase that would provide a convenient handle for his ideas on the mutability of species, though the words *transmutation* and *transformism* are often used to characterize his views.[25] The debate over species had yet to adopt permanent vocabulary.

The question may fairly be asked why Lamarck rather than Erasmus Darwin receives primary credit as the first evolutionist. Erasmus Darwin did present his ideas first; yet he cast his views in verse and, though his work was known, he did not have students or followers who adopted his views in any conventional sense. In contrast, Lamarck presented his ideas in prose. He also gave lectures that attracted many receptive listeners.[26] It should also be noted that in late eighteenth-century and early nineteenth-century Europe there were numerous individuals who speculated in various ways about the mutability of species.[27]

Cuvier's approach to species was anti-Lamarckian. He did not believe that species were mutable. His approach to species was largely drawn from his expertise as a comparative anatomist who specialized in vertebrates. Cuvier's interpretation of anatomy was largely functionalist. He asked the question, "What is the correlation between the structure of an animal and its manner of life?" He believed that the skeletons of the various animals he studied were so well adapted to the particular circumstances of their existence that no slight Lamarckian adjustment in their body design was possible. The fit between organism and circumstance was too exact.

Another approach was taken by Geoffroy Saint-Hilaire (1772–1844), also a member of the Muséum elite. His tack was to ask the question, "Can the organization of vertebrate animals be reduced to a uniform type?"[28] His approach emphasized the study of anatomical form and structure (morphology). In principle, such an approach could serve either an evolutionary or a nonevolutionary perspective. An evolutionary approach to morphology would emphasize similarities in structure between animals of different groups, while an antievolutionary approach would emphasize difficulties in adequately explaining how such similarities had arrived. In balance, Saint-Hilaire emphasized similarities and tacitly affirmed past evolutionary connections.

Since the Muséum d'Histoire Naturelle was the most substantial institution of its kind in Europe, it was visited frequently by naturalists born outside France. Several people prominent in the education and career of Charles Darwin made the journey to Paris during the 1820s and 1830s. Among them was Edward Grant (1793–1874), who taught Darwin at the University of Edinburgh from 1825 to 1826. In his autobiography Darwin recalled of Grant, "He one day, when we were walking together burst forth in high admiration of Lamarck and his views on evolution. I listened in silent astonishment."[29] From this we know that Charles Darwin was exposed to Lamarck's ideas even as a young student.

Another important visitor to Paris in the 1820s was British geologist Charles Lyell (1797–1875). As a young man Lyell had trained for the law, but, after beginning to practice, found himself drawn increasingly to the study of geology. His ability to argue a case served him well: It was he who framed for Charles Darwin (and for others) what we may now term the "question" of evolution. To explain what is meant by this statement, we need to refer to Lyell's own history. While a student at the University of Oxford he had benefited from the lectures of geologist and paleontologist William Buckland (1784–1856), who followed Cuvier on most points. Lyell was also influenced by the ideas of James Hutton (1726–1797), the Scottish geologist who emphasized the great age of the earth and the ongoing activity of geological forces within it. In the late 1820s Lyell visited Paris, where he was entertained by Cuvier and introduced to a broad range of ideas on species, including those of Lamarck. His vision dazzled by the perspectives offered by his British and French colleagues, Lyell conceived the plan of writing an introductory book on geology that would appeal both to the public and to the specialist. Part of the reason he wanted to span such a wide audience was practical: A successful author could sustain his own career. At the time there were few career opportunities open to geologists, zoologists, and botanists, and those who were interested in these subjects generally had to support themselves from their own resources.

Lyell published his book *The Principles of Geology* in three volumes.[30] In the first volume he argued against attempting to correlate the earth's history with the biblical flood, as his mentor Buckland had done. He also argued that geological causes presently acting should be used to explain thc former changes on the globe. In the second volume he considered species. In the third volume he applied his general philosophical outlook to the concrete details of the geological record: strata, formations, fossils.

Lyell's first volume arrived on young Charles Darwin's desk at an opportune time—1831—just as he was finishing his undergraduate degree and preparing for his life's work. Later, Darwin would say of Lyell's book:

> I have always thought that the great merit of the Principles, was that it altered the whole tone of one's mind & therefore that when seeing a thing never seen by Lyell, one yet saw it partially through his eyes.[31]

On the purely geological side Lyell influenced Darwin toward uniformitarianism, a term coined by philosopher William Whewell (1794–1866) to represent a presumption in favor of a constancy in geological forces over time. Lyell was particularly strong in emphasizing the great age of the earth, although there was, in his time, no means to assign an absolute number to its age. Lyell's influence on Darwin regarding species was of paramount importance. Like other geologists of his generation, Lyell accepted as fact Cuvier's claim that some species had become extinct over the course of the earth's history. However, Lyell differed from many of his peers in two ways. He believed that species became extinct gradually as numbers of individuals declined owing to corresponding gradual changes in physical circumstances such as climate; he did not believe that geological catastrophes had destroyed whole batches of species. Further, he posited that new species were introduced gradually to replace those that had been lost.

In his second volume Lyell came down strongly against Lamarck, although he admired much about his approach (Document 10). In 1827, on first reading Lamarck, he had remarked in a letter to his fellow geologist Gideon Mantell (1790–1852), "I devoured Lamarck en voyage. . . . His theories delighted me more than any novel I ever read, and much in the same way, for they address themselves to the imagination, at least of geologists who know that mighty inferences which would be deducible were they established by observations." Lyell further praised Lamarck for courageously admitting that "his argument . . . would prove that men may have come from the Ourang-Outang [orangutan]." But then, in a more reflective mood, Lyell wrote, "But after all, what changes species may really undergo! How impossible will it be to distinguish and lay down a line, beyond which some of the so-called extinct species have never passed into recent ones." Of Lamarck altogether he noted that "I read him rather as I hear an advocate on the wrong side, to know what can be made of the case in good hands."[32]

Lyell's use of the word *advocate* is instructive. He thought of Lamarck as debating one side of the question while he was debating the other side. This was a debating society or, given Lyell's training, a courtroom. He took Lamarck's rather rambling presentation in *Philosophie zoologique* and turned it into a question: Do species have a real and permanent existence in nature, and are they capable of being modified indefinitely? Once the subject of species mutability was framed as a question, it invited answers. It also invited speculation as to what astronomer John Herschel (1792–1871) referred to as "that mystery of mysteries the replacement of extinct species by others" (Document 11).

CHARLES DARWIN AND THE EVOLUTION DEBATE

With the question of evolution on the table, we must now turn to discuss the role of Charles Darwin in its solution. Lyell and Darwin shared a common background. At a time when less than 1 percent of the population attended university (and this group all male), both men were university graduates—Lyell from the University of Oxford, Darwin from the University of Cambridge. Of the three traditional professions (law, medicine, theology), Lyell had been intended for law, Darwin first for medicine (when he was at Edinburgh) and then for the church (when at Cambridge). In the end both young men deviated from the courses laid out for them by their fathers, becoming gentlemen scientists, whose careers were supported by family money and supplemented by earnings from their publications. Their professional competence was assured by various scientific societies that set standards for research. Of these societies the most prestigious was the Royal Society of London, founded in 1660. Both Lyell and Darwin were Fellows of the Royal Society. Among the smaller societies, the one especially important to both of them was the Geological Society of London. Yet there was one important difference in their background. Charles Darwin's grandfather Erasmus had been an early advocate of evolutionary ideas. That circumstance of birth made Darwin more open to this new idea of evolution (transmutation) than was Lyell.

Charles Darwin's life was both adventurous and ordered, dangerous and serene. It spanned much of the nineteenth century, like that of Queen Victoria (1819–1901) who gave her name to the era. His own life was of a piece with the century, for he benefited from the prosperity of Victorian Britain, and its ambitions to dominate world events through its command of the seas. The grandson of two members of the Lunar Society (Josiah Wedgwood and Erasmus Darwin), Charles Darwin carried forward its ideals—liberal in politics and religion, imaginative in science, and industrious in daily life. Because of the fine education allowed to him by his father's wealth, and his good fortune in being born in a time of relative peace, Darwin was able to chart his own course in life.

At the University of Cambridge he studied with John Stevens Henslow (1796–1861), who was both a geologist and a botanist. While not a transmutationist, Henslow was interested in studying variation within species.[33] He trained Darwin in state-of-the-art techniques of collecting and preserving specimens. He also shaped Darwin's career: Had Darwin remained in England, he would likely have followed Henslow's path in combining a clerical appointment with the pursuit of natural history. (To do so, Darwin would not have had to attend university beyond his baccalaureate degree; there was no degree course in divinity.) But, as it turned out, Henslow shaped Darwin's purely scientific career as well. It was he who in late summer 1831 recommended Darwin for service as naturalist aboard HMS *Beagle*. This voyage, which lasted nearly five years, transformed Darwin's life.

THE VOYAGE OF HMS *BEAGLE*, 1831–1836

HMS *Beagle* was a surveying ship sent out by the British Admiralty primarily to chart the coastline of the southern portion of South America and secondarily to execute a circumnavigation of the globe. Darwin was proud to be a part of the tradition of exploration represented by the voyage, and there is a tone of patriotism that permeates his correspondence from the trip. The route of the 1831–1836 *Beagle* voyage is shown in the map on page 18. The voyage was an enormous undertaking. While the hardships of travel by sea were considerable, the rewards were correspondingly great. Darwin and his shipmates saw much of the world, including societies far different from their own, like that of the canoe Indians of Tierra del Fuego. While in Brazil, Darwin encountered slavery firsthand, and like his grandfathers, he found it repugnant. Slavery was abolished in British colonies in 1833 but remained in Brazil as it did in the United States. More positively, though, what Darwin gained from the voyage was a rich variety of experiences that endured in his memory. He recorded those memories in his narrative account of the voyage: *Journal of Researches into the Geology and Natural History of the Various Countries Visited by H.M.S. Beagle*, published in 1839 (Document 12).

On the scientific side, as the prospect of a voyage approached, Henslow, while allowing that Darwin was not a "finished Naturalist," charged him with "collecting, observing, & noting any thing worthy to be noted in Natural History."[34] During the voyage Darwin did all these things extremely well; as a member of the rural gentry he knew he was a good shot, and he knew his way around the outdoors. Moreover, he had with him on the *Beagle* a fine library of reference works. While ultimate judgments on their identity had to be deferred until he consulted experts at home, he could provide tentative identifications for specimens during the voyage.

Among the several thousand specimens he collected, two groups of zoological specimens were of particular importance to him as he came to embrace the idea of species transmutation.[35] The first group included organic remains of extinct mammals. Ever since the excitement over *Megatherium*, researchers in Europe and the Americas had desired more physical evidence of extinction. In 1832, during the first year of the voyage, Darwin had the opportunity to collect in the same area (now Argentina) where *Megatherium* had originally been found. Greatly aided by local people who guided him to specimens, he found more samples of *Megatherium* bones and immediately dispatched them to England, where they were well received. He also found bones of unknown origin. One particularly interesting specimen was a collection of scutes (external bony plates) that reminded him of large-scale versions of the tesselated bony armor that composed the carapace of modern-day armadillos inhabiting the very area he was exploring. At the time Darwin was on the *Beagle*, such scutes were thought by geologist Buckland to have been the hide of the *Megatherium*. Later, based on a more complete specimen of *Megatherium*, Richard Owen (1804–1892), a comparative anatomist, reinterpreted the scutes as the remains of what he termed a "Gigantic Extinct Armadillo," or *Glyptodon clavipes* (Document 13). For Darwin, after the voyage, it was a short step to see the *Glyptodon* and the armadillo as related by descent: change over time.

The second key group of specimens were those from the Galápagos Islands, which Darwin visited in 1835. By the time the *Beagle* reached the islands on its surveying mission, Darwin was a thoroughly experienced field naturalist and well acquainted with South American fauna and flora. He was prepared to recognize the subtle characteristics of the archipelago's species. As he later put it in 1839 in his published account of the *Beagle* voyage, "The natural history of this archipelago is very remarkable: it seems to be a little world within itself; the greater number of its inhabitants, both vegetable and animal, being found nowhere else."[36] What struck him was that some species showed slight variations

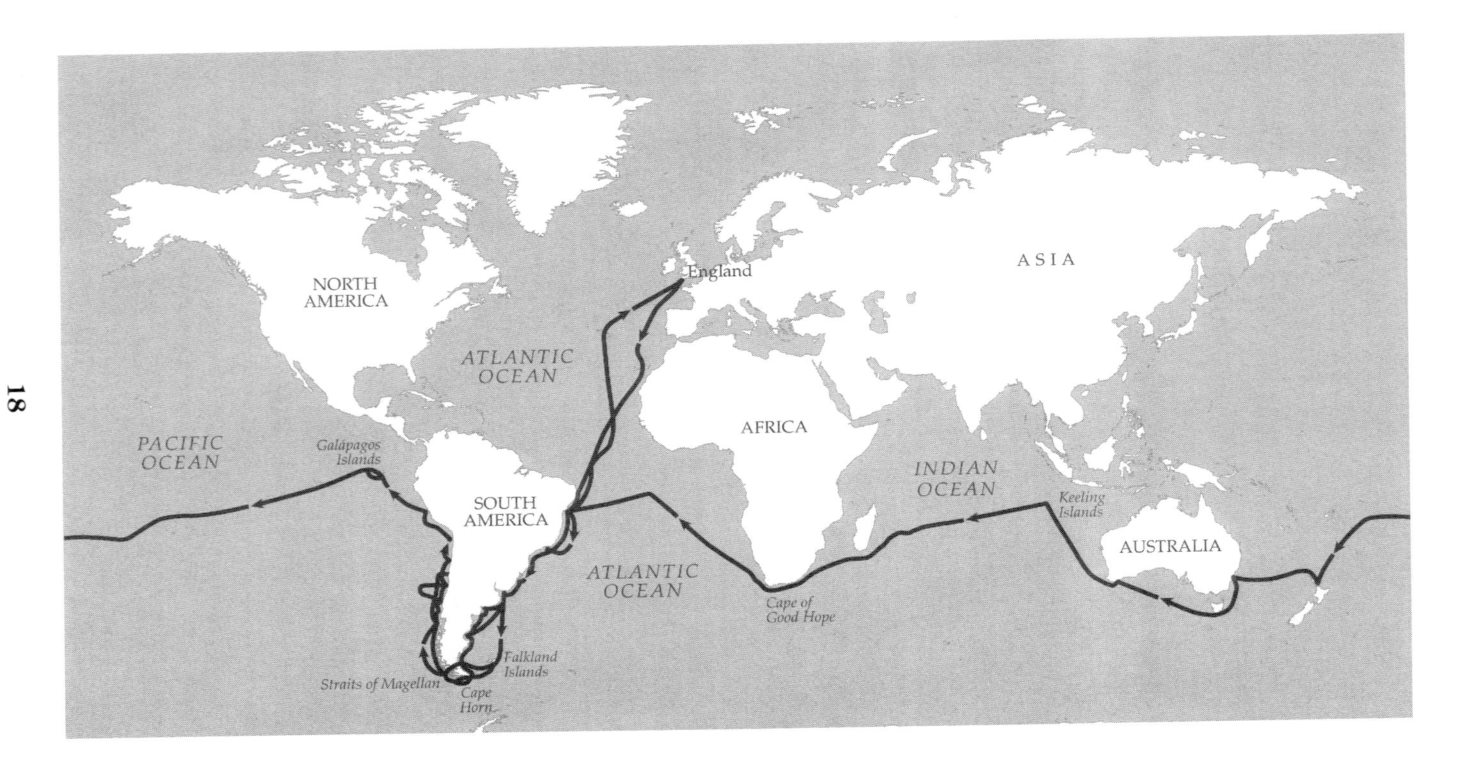

NORTH AMERICA
ATLANTIC OCEAN
England
ASIA
AFRICA
PACIFIC OCEAN
Galápagos Islands
SOUTH AMERICA
INDIAN OCEAN
Keeling Islands
AUSTRALIA
ATLANTIC OCEAN
Cape of Good Hope
Falkland Islands
Straits of Magellan
Cape Horn

from island to island. The mockingbirds displayed this characteristic most strikingly. In notes written during the voyage, he identified the mockingbirds by their common name—"thenca"—and speculated about their unusual character (Document 14). On Darwin's arrival back in England with the specimens, at a meeting of the Zoological Society of London on February 28, 1837, ornithologist John Gould (1804–1881) declared Darwin's mockingbird varieties to be good species. From that point on, Darwin believed that he had witnessed the creation of new species. This insight occurred in the spring of 1837. Thereafter he began to imagine what might be the case "if one species does change into another."[37] He had become a transmutationist. See Figure 1 for illustrations of two of the Galápagos mockingbirds.

DARWIN'S DEVELOPMENT OF A THEORY

The example of the Galápagos mockingbirds gave Darwin a model for how new species might come into existence. During the summer of 1837 he opened a notebook to pursue the whole subject of transmutation further. Interestingly he began his deeper inquiry by first rereading his grandfather Erasmus Darwin's *Zoonomia*, an early transmutationist work; he headed the first page of his notebook "Zoonomia" and went on

(Opposite) Map 1. *Route of HMS* Beagle, *1831–1836.*
This map (or, being nautical, chart) shows the route of HMS *Beagle*, which sailed from England on December 27, 1831. The ship was under the command of Robert FitzRoy (1805–1865), a career naval officer and expert in naval surveying (hydrography) and in meteorology. On the outbound journey, the *Beagle* crossed the Atlantic Ocean heading for the southern portion of South America, where its chief assignment was to survey the coastlines of that area (hence the numerous track marks in the area of Cape Horn at the tip of South America). On the return journey, the *Beagle* sailed from the west coast of South America across the Pacific Ocean, making what was for Charles Darwin a fateful stop at the Galápagos Islands in September and October 1835. Eventually, the *Beagle* completed its circumnavigation, arriving back in England on October 2, 1836. For a more detailed presentation, see "General Chart shewing the Principal Tracks of H.M.S. Beagle" in Robert FitzRoy, ed., *Narrative of the Surveying Voyages of His Majesty's Ships Adventure and Beagle*, appendix to vol. 2 (London: Henry Colburn, 1839), frontispiece (available on www.darwin-online.org.uk). The chart shown here is modeled on the simplified version in Frederick Burkhardt, ed., *Charles Darwin: The "Beagle" Letters* (Cambridge, U.K.: Cambridge University Press, 2008), flyleaf.

to discuss his grandfather's views on reproduction.[38] In this notebook (which Darwin labeled "B") he also recorded other examples of the phenomena of geographic isolation being correlated with new species. He wrote: "According to this view animals, on separate islands, ought to become different if kept long enough—apart, with slightly differen[t] circumstances.—Now Galapagos Tortoises, Mocking birds; Falkland Fox—Chiloe, fox,—Inglish & Irish Hare."[39] This brief entry illustrates something of the speculative content of Darwin's developing vision. A longer extract from this same notebook contains more of Darwin's transmutationist speculations and illustrates his free and unpolished style (Document 15).

Once Darwin had committed himself to a belief in the mutability of species, he had more work ahead of him. Clearly he needed to tackle the subject of the origin of man. He quickly came to believe that human beings had a natural origin, that as a species they were descended from earlier forms. In Darwin's emerging view, humans were animals. He knew this would be a troublesome view to some, and he recollected the condescension of the slaveholder toward the slave to make his point: "Animals—whom we have made our slaves we do not like to consider our equals.—Do not slave holders wish to make the black man other kind?" But, he continued, "if we choose to let conjecture run wild, then animals our fellow brethren in pain, disease death & suffering & famine; our slaves in the most laborious work, our companion in our amusements, they may partake, from our origin in one common ancestor[.] [W]e may be all netted together."[40]

(Opposite and p. 22) Figure 1. *Galápagos Mockingbirds.*
Charles Darwin noted differences in the appearance of mockingbirds among islands in the Galápagos. Illustrations of two of the birds he collected are shown. These illustrations are taken from *The Zoology of the Voyage of H.M.S. Beagle:* Part III, *Birds*, edited by John Gould (London: Smith, Elder, and Co, 1841). The mockingbird on the left is *Mimus trifasciatus* (now called *Nesomimus trifasciatus*), from Charles Island (Isla Floreana). The bird on the right is *Mimus melanotis* (now called *Nesomimus melanotis*), from Chatham Island (Isla San Cristóbal). These were the first two mockingbirds Darwin saw on the islands, and he was impressed that he could easily distinguish between them. The Charles Island mockingbirds were larger and with noticeable patches of black feathers across the top of the chest. The Chatham Island mockingbirds were smaller and paler but with a distinct pattern of dark feathers behind their eyes.

Mimus trifasciatus.

Mimus melanotis.

As an evolutionist, Darwin also began to think about religion. Most generally his tack in his theoretical notebooks from just after the voyage was to suggest that it was a simpler and more "sublime" view to regard God as acting through "fixed laws of generation."[41] Sometimes he pushed traditional belief harder, at one point in his notebooks commenting on himself "oh you Materialist!"[42] Of course these were private entries in a private notebook; only those close to him would have known his thoughts. One person who did know his thoughts was Emma Wedgwood (1808–1896), his first cousin and confidant, who became his wife in January 1839. Shortly after their marriage she expressed her reservations as to the direction of his thought, asking, "May not the habit in scientific pursuits of believing nothing till it is proved, influence your mind too much in other things which cannot be proved in the same way, & which if true are likely to be above our comprehension" (Document 16). Their difference over religion proved to be an ongoing one in their marriage, Emma continuing to attend weekly services at their local parish church, Charles remaining more frequently at home. In later life Darwin wrote to his friend Joseph Hooker (1817–1911), "My theology is a simple muddle: I cannot look at the Universe as the result of blind chance, yet I can see no evidence of beneficent design, or indeed of design of any kind in the details."[43] At his death Charles Darwin received a religious funeral and was buried in Westminster Abbey in London.

Man's place in nature, the relationship of science and religion—these were large cosmological questions. More pressingly, in the late 1830s, as he was keeping his transmutation notebooks, Darwin had to face a scientific question: How were adaptations to be explained? The exquisite manner in which the eye worked (to take only the most classic example) was a traditional argument used in support of the existence of design in nature, and therefore of a designer. When he opened his notebooks in 1837, Darwin had no explanation, but in the course of reading and thinking about the subject he discovered one. He called his explanation "natural selection." Darwin was inspired not only by the work of his grandfather Erasmus Darwin, but also by Erasmus's contemporary Thomas Malthus, who theorized that human progress was limited by a natural tendency toward overpopulation. Darwin recorded his first reading of Malthus on September 28, 1838, with excitement. He was impressed with Malthus's argument that population, when unchecked, goes on doubling itself every twenty-five years, or increases in a geometric ratio.

From that statement Darwin drew the conclusion that the "final cause" of this process "must be to sort out [the] proper structure & adapt it to change."[44] Given an excess of individuals, those with a variation that was in some manner beneficial would survive. As Darwin put it, "Suppose six puppies are born & it so chances, that one out of every hundred litters is born with long legs & in the Malthusian rush for life, only two of them live to breed, if circumstances determine that, the long legged one shall rather oftener than any other one survive in ten thousand years the long legged race will get the upper hand."[45]

By the end of 1838 Darwin had the essential elements in place for what would become his theory of evolution through natural selection. What he lacked was some demonstration of this process actually occurring. He believed that the process in nature would occur slowly, too slowly for direct observation. He therefore addressed himself to the study of domesticated productions. For example, in 1839 he surveyed the opinions of practical men in his privately printed "Questions about the Breeding of Animals."[46] The material that he gathered on what he called "artificial selection"—a phrase he used as a complement to "natural selection"—took prominent place in all of his subject work on the species question. Thus the first chapter of *On the Origin of Species* would eventually be entitled "Variation under Domestication."

The year 1839 was a milestone for Darwin. He turned thirty years old, he married, and his account of the voyage—the *Journal of Researches*—was published. Further, he had already in mind the outlines of his theory of evolution through natural selection. Yet there was an immediate obstacle to Darwin's success in the scientific world. In 1837 Louis Agassiz (1807–1873), a Swiss zoologist and former student of Cuvier's, had proposed the notion of glacial ages. In Agassiz's view, species were destroyed at the end of each age.[47] If this were so, Darwin's theory of descent would be impossible. Initially Darwin reacted strongly against Agassiz's glacial hypothesis, but eventually accepted a modified version of it. In Darwin's interpretation of Agassiz, the catastrophes caused by glacial periods did not destroy all living species.

We will now turn to look briefly at the next period in Darwin's life. Until 1846, when *Geological Observations on South America* was published, Darwin was busy wrapping up his obligations as naturalist to the voyage.[48] He was quite aware that, as his mentor Henslow had foretold, it had taken him twice as long to publish his observations as to make them. During the years since the voyage he had been equally busy, at a more private level, with developing his ideas on transmutation. In 1842 he wrote up a brief "sketch" of his ideas. Two years later he wrote a much longer and more formal treatment. These early essays were written "for the drawer" (that is, for his own private contemplation) and were not published until after his death; however, intellectually they were the basis of the book he later published as *On the Origin of Species.*[49] While Darwin was not yet ready to publish in the 1840s, he did write a testamentary letter to his wife Emma, asking that she arrange for publication of the essay should he die prematurely.[50]

At this juncture, in the mid-1840s, Darwin faced the question of what to do next. His health was not robust, so he could not have pursued any scientific activity that involved travel. He then made an interesting choice of research topics. He was aware that most natural historians devoted a great portion of their time to surveying and classifying one or another group of organisms. Taxonomy was at the heart of the discipline. He had not yet done such work. He was also aware that he would increase his stature among zoologists were he to undertake such a task. This would eventually aid the reception of his theoretical work on the species question. His friend, botanist Joseph Hooker (1817–1911), affirmed the wisdom of this choice of direction. The group Darwin chose to study were the barnacles, or cirripedes, the small marine crustaceans familiar to sailors worldwide. He had noted unusual features regarding the reproductive structures of members of the group, which he wanted to study further. His initial rather narrow interest came to encompass the entire group, and over the course of eight years, he produced a standard classification of both living and fossil cirripedes.

What of Darwin's British peers in the field of natural history? Comparative anatomist Richard Owen was among the most prominent. His larger-scale interest was in developing ideas about what he termed the vertebral archetype. This was quite a different line of inquiry from Darwin's. John Gould, the ornithologist, gave Darwin a mine of information but remained convinced of the stability of species. Still, lines of evidence continued to be developed that pointed toward evolution. One strong line of evidence was the fossil record. The testimony of geologist Roderick Murchison (1792–1871) was significant on this point (Document 17).

Of greatest importance, there was one man ready in Victorian Britain in the 1840s to draw an evolutionary conclusion from the increasing detailed knowledge of progression in the fossil record. That man was

Robert Chambers (1802–1871). A successful publisher in Edinburgh, Chambers was an Erasmus Darwin for his generation in the sense that he promulgated an evolutionary philosophy. There were two important differences in the facts of their publication. First, Chambers published his views anonymously. The second difference was that the nature of publishing had changed between the late eighteenth century and the mid-nineteenth century. Where Erasmus Darwin's works were produced for a small market, Chambers's *Vestiges of the Natural History of Creation* was produced for a large middle-class reading public—what economists would now call a mass market. *Vestiges* went through eleven editions from 1844 through 1860, and it had at least one hundred thousand readers.[51] The book was, in the words of historian James Secord, a Victorian sensation:

> As readable as a romance, based on the latest findings of science, *Vestiges* was an evolutionary epic that ranged from the formation of the solar system to reflections on the destiny of the human race. . . . In a hugely ambitious synthesis, it combined astronomy, geology, physiology, psychology, anthropology, and theology in a general theory of creation. It suggested that the planets had originated in a blazing Fire-mist, that life could be created in the laboratory, that humans had evolved from apes.[52]

While the anonymously published *Vestiges of the Natural History of Creation* did not appeal to most scientists in that its larger claims were unproven, the inspirational quality of its sweeping vision familiarized the Victorian reading public with evolutionary ideas, or what Chambers referred to as the "hypothesis of the development of the vegetable and animal kingdoms."[53] This popular—even sensational—book introduced evolutionary ideas to a new generation and paved the way for the relatively smooth public acceptance in Britain of Darwin's *Origin* fifteen years later (Document 18).

Among the many readers for whom *Vestiges of the Natural History of Creation* was an important book was Alfred Russel Wallace (1823–1913). As a young man of wide interests but without the financial means to attend university, Wallace was an ideal reader for the anonymous author of *Vestiges*. In fact Chambers and Wallace were rather similar in their adventurous intellectual outlook but straitened circumstances. Of course Wallace read other authors (among them Malthus, Humboldt, and Lyell) and knew Lamarck's views. He was also familiar with Charles Darwin's account of his travels aboard the *Beagle*, and he had the benefit of access to the second edition of 1845, where Darwin, while not writing explicitly as a transmutationist, was more revealing of his theoretical leanings than he had been in the first edition of 1839. In addition to becoming widely read in the literature of current science, Wallace also developed a strong interest in natural history, enhanced by his friendship with entomologist Henry Walter Bates (1825–1892). Inspired by a desire to travel, Wallace proposed to Bates that they go to South America to collect exotic specimens, whose sale would provide their livelihood. Wallace explored the Amazon basin from 1848 to 1850 and became a published expert on the geographic distribution of plants and animals in the region.[54] From 1854 through 1862 he traveled in the Malay Archipelago (now Indonesia and Malaysia). This journey also resulted in a scientific travel narrative of lasting importance.[55]

Like his contemporary Charles Darwin, Wallace was theoretically inclined. In 1855, writing from Borneo, he published an article in a London journal that drew the admiring attention of Charles Lyell. The article was entitled "On the Law Which Has Regulated the Introduction of New Species" (Document 19). Wallace's primary conclusion was that "every species has come into existence coincident both in space and time with a pre-existing closely allied species." To a present-day reader who begins from a presumption of the descent of species, Wallace's conclusion may seem like a commonplace, but for Lyell it was a revelation, snapping into focus an obvious conclusion that had lain just under the surface of his previous thought. Wallace's 1855 article did not strike Darwin as original, for he had already reached an identical conclusion years earlier. But Wallace's next theoretical contribution was a bombshell. In mid-1858 Darwin received a letter from Wallace containing an article that stated ideas quite similar to Darwin's notion of evolution through natural selection. Wallace requested that, if Darwin were agreeable, he would show the article to Lyell. What happened next forced the creation of the book known to the world today as *On the Origin of Species.*

THE BIRTH OF *ON THE ORIGIN OF SPECIES*

In the mid-1850s, once he had finished work on classifying living and fossil barnacles, Darwin had turned to writing up his theory of evolution. It was to be a large book. The tentative title was "Natural Selection." In the course of writing this draft, Darwin began to confide more openly in several peers regarding the content of his work.[56] Formerly when he spoke of his theory to intimates he had described his understanding of the descent or genealogy of species; now he spoke also of selection as the agency for change.

While Lyell continued to be a key confidant in the late 1840s and early 1850s, Darwin increasingly turned to botanist Joseph Hooker who, as it happened, was also a son-in-law of his beloved Cambridge mentor, Henslow. Asa Gray (1810–1888), an American botanist who was both a professional colleague and a friend of Hooker's, was also taken into Darwin's confidence. As was characteristic of his correspondence, Darwin exchanged his own ideas for information, in the case of Gray regarding the geographic distribution of American plant species. In an 1857 letter to Gray, Darwin went so far as to describe in fairly concrete detail his theory of evolution through natural selection (Document 20). The American naturalist was thus positioned to play a role in the drama that followed.

Unsure of what to do with Wallace's 1858 manuscript, Darwin passed the problem on to Lyell and Hooker. They decided that Wallace's article should be published alongside a selection from Darwin's 1844 essay on his theory (never before published, though Hooker had read it previously) and the summary of the theory Darwin had enclosed in his 1857 letter to Gray. All three of these documents were published with an accompanying letter by Lyell and Hooker in the *Journal of the Linnean Society of London*.[57] The first public announcement of a theory of evolution through natural selection was thus a joint production of Charles Darwin and Alfred Russel Wallace. The relationship between Darwin and Wallace is one of the more interesting ones in the history of science since each followed a parallel course in coming in some ways independently to similar conclusions. On this point compare their own autobiographical accounts of their discoveries (Documents 21 and 22). Throughout his life Wallace deferred to Darwin on the priority question, that is, who first devised the theory of evolution through natural selection.

Once the cat was out of the bag regarding the theory, Darwin's plan for his *Natural Selection* manuscript clearly required rethinking. He decided to set aside his longer work in favor of a shorter publication that would address the public more directly. In the course of eleven months he wrote the manuscript that was intended as an "abstract" of his views. Still, the novelty of his views, as well as the form in which they were expressed, gave pause to the intended publisher of the work, John Murray (1808–1892). Murray sent the manuscript out for a reading to a trusted confidant, Whitwell Elwin (1816–1900) (Document 23). Despite Elwin's doubts, Murray published the *Origin* as Darwin had written it.

On the Origin of Species was organized in a straightforward manner (Document 24). The five opening chapters, only 130 pages in length, carry the nub of the argument. Darwin began by establishing that the well-known successes of domestic breeding showed that there was a sufficient fund of variation among individuals for selection to take place (Chapter I: "Variation under Domestication"). He then argued, by analogy, that a similar fund of variation must exist in nature on which selection could act (Chapter II: "Variation under Nature"). Next Darwin argued the Malthusian case that more individuals were produced in the ordinary course of generation than could survive (Chapter III: "Struggle for Existence"). As capstone to his argument, Darwin concluded that nature exercised a more powerful role than man in determining which individuals and species do survive (Chapter IV: "Natural Selection"). Darwin supported his argument by noting that ill-adapted individuals left fewer descendants than well-adapted individuals. From this he concluded that a species with insufficient numbers of individuals to sustain itself would become extinct. Darwin devoted the remainder of his book to a variety of topics, including one chapter that candidly discussed "Difficulties of the Theory." Lest his reader be too discouraged by what remained unknown, he concluded his book with a chapter, excerpted in Document 24, summarizing his conclusions and inviting his reader to consider that "there is grandeur in this view of life" in which "from so simple a beginning endless forms most beautiful and most wonderful have been, and are being, evolved."

RESPONSE TO *ON THE ORIGIN OF SPECIES*

Darwin's *Origin of Species* was reviewed extensively in Britain and elsewhere.[58] Language and tradition affected the immediate response to the *Origin*. Since the book was written in English, response occurred more rapidly in English-speaking countries. In areas of the world where translation of the book was required, response necessarily came more slowly. Within Darwin's lifetime the book was translated into eleven European languages: German (1860), Dutch (1860), French (1862),

Russian (1864), Italian (1864), Swedish (1869), Danish (1872), Polish (1873), Hungarian (1873–1874), Spanish (1877), and Serbian (1878). Since Darwin's death, the *Origin* has been translated into an additional thirty-three languages, including Chinese and Arabic.[59]

The nature of the scientific tradition already in place within a country also shaped how Darwin's *Origin* would be received. The different treatment that the *Origin* received in German-speaking lands (Germany was not yet a unified country) and in France illustrates this point. German universities had developed a strong tradition in biological research during the nineteenth century, with particular emphasis on microscopic work and cell biology. For German biologists, Darwin's more field-based studies seemed complementary to their own work. Understandably, then, when a German translation of the *Origin* appeared in 1860, the response to it among German scientists was largely positive. Preexisting scientific tradition could also work against the book. In France the antitransmutationist views of Cuvier were still strong, so strong that no prominent scientific author stepped forward to translate the book into French. Thus it was that the first French translation of the *Origin* was done by Clémence Royer (1830–1902), a woman who valued Darwin's work more for its broad philosophical views than for its contribution to answering narrow scientific questions.[60]

In Britain itself, response to the book was immediate. On November 24, 1859, *On the Origin of Species by Means of Natural Selection* was published by John Murray in a print run of 1,250 copies, all of which were sold that day to the booksellers' trade or retained for review or the author's use. (Of the initial print run, 500 copies were bought by Mudie's Subscription Library, which was a commercial lending library; this meant that each copy of the book had many readers.) A new edition was called for immediately; in the end, the *Origin* would go through six editions, and it has never been out of print since its initial publication. Its wide readership was owing to the inherent interest in the subject as well as to Darwin's talent for expressing complicated ideas clearly.

The book was widely commented on. In nineteenth-century Britain there were numerous general-interest periodicals in which works such as the *Origin* were reviewed.[61] Thus the book and its author soon became household words in Britain. Since Darwin, an invalid since the early 1840s, rarely left home, he did not champion his own work in public. That task devolved on several early supporters, most prominently

Thomas Henry Huxley (1825–1895), who was then a professor of natural history at the Royal School of Mines. A famous clash over the *Origin* occurred at the 1860 meeting of the British Association for the Advancement of Science, a broad-based organization formed in 1831 to promote science in Britain. The meeting venue for that year was the University of Oxford. Several at the meeting objected to a theory that would make man descended from an ape. Huxley promptly rose to defend Darwin and to accept the possibility that humans were descended from apes (Document 25). As the Oxford debate over the *Origin* illustrates, public debate over the question of the mutability of species quickly moved to a discussion of human ancestry.[62]

Reception of the *Origin* in the United States mirrored that in Britain in the sense that early supporters of the theory fairly quickly persuaded the majority of scientists that the theory had merit. (It was not until the Scopes Trial of 1925, which centered on the teaching of evolution in the public schools, that the theory became widely controversial in the United States.[63]) The early positive response in the United States was led by Darwin's longtime confidant, Harvard University botanist Asa Gray. Gray helped to arrange for an American edition of the *Origin* and promoted the book to colleagues.

There was, however, tension among the Harvard faculty over the *Origin.* Harvard zoologist Louis Agassiz, a student of Cuvier's and a well-known critic of transmutationism, opposed Darwin. Gray was eager to take on the responsibility for defending the new theory and was aware that he would be arguing against Agassiz in attempting to persuade an educated American audience of the merits of the *Origin.* In a review for the March 1860 *American Journal of Science and Arts*, Gray praised Darwin's work, argued against Agassiz's known positions, and also—and this perhaps more surprising—urged that Darwin's theory was consistent within a traditional natural theological framework (Document 26).[64] Agassiz's comments on the *Origin*, published later in the year in the same journal, were harsh. He did not regard Darwin as having supplied the facts to establish his case (Document 27). Since Gray and Agassiz were colleagues, their difference of opinion was felt close at hand and would continue to resonant after their deaths (Document 28).

While Cambridge, Massachusetts, was the scene of the Gray-Agassiz clash, other intellectual centers in the United States also shared in the debate. As a repository for natural history collections, the Smithsonian

Institution in Washington, D.C., was inevitably drawn into the debate over evolution. Joseph Henry (1797–1878), the Smithsonian's secretary, was favorably disposed toward Darwin's theory. Although a physicist, Henry had known of Darwin's abilities as a naturalist since 1837 when he met John Stevens Henslow while on a visit to the University of Cambridge. At that time Henslow, his host, presented him with a copy of some of Darwin's letters from the *Beagle* voyage.[65] Of even more importance, Henry was a close friend of Asa Gray and through Gray knew of Darwin's conclusions in the *Origin* even before publication of the book. Influenced by his knowledge of Darwin's evolutionary thesis, and in consideration of his own knowledge of the vast changes constantly at work on the earth's surface, Henry wrote in his "lock box" (diary) in 1862, "Now the important deduction which we can draw from this is that creation is, as it were, still going on."[66] Guided by Gray, Henry had joined the evolutionist camp.

While Henry was corresponding with Asa Gray over evolution, he was also serving the federal government as a scientific adviser. Abraham Lincoln (1809–1865) was elected president in 1860; by April 1861 the nation was at war. The issue of the war was slavery, narrowly whether it should be prohibited in the Western territories, more generally whether as an institution it was destined for what was termed at the time "ultimate extinction."[67] Henry assisted Lincoln during the war. He also continued his work in science, corresponding with Gray on both scientific and political matters. Similarly Gray continued his work in science and expressed his views on politics as he wrote to various people.

Gray was both an evolutionist and an abolitionist. In Darwin he found a sympathetic correspondent on political as well as scientific subjects (Document 29). What one realizes from their correspondence is how interlocking the issues were—the subjects of slavery and of evolution are intertwined in their letters. In his letters to Darwin, Gray reported on progress on the Union side. When Darwin expressed concern that Lincoln was not moving to end slavery with sufficient dispatch, Gray defended the president as "our representative man." Darwin and Gray were eager to see slavery abolished in the United States. Equally, both men were keen to advance the cause of evolutionary ideas. There were differences between the two men. Gray, the practicing Presbyterian, sought to retain as part of his evolutionary views the argument for the existence of God from the evidence of design in nature. While not following him on that point, Darwin respected Gray's views and sought to publicize them. Darwin did not wish his theory of evolution to be bound to antireligious sentiment.

By the mid-1860s Darwin's high place as a scientist was recognized with the Copley Medal, the most prestigious honor given by the Royal Society of London. In nominating Darwin for the award, paleontologist Hugh Falconer (1808–1865) praised him as one of the "Great Naturalists of all Countries and of all time" and wrote this about his theory of evolution through natural selection:

> This solemn and mysterious subject had been either so lightly or so grotesquely treated before, that it was hardly regarded as being within the bounds of legitimate philosophical investigation. Mr. Darwin after 20 years of the closest study and research, published his views, and it is sufficient to say that they instantly fixed the attention of mankind throughout the civilized world. The efforts of a single mind should have arrived at success on a subject of such vast scope, and encompassed with such difficulties was more than could have been reasonably expected—and I am far from thinking that Charles Darwin has made out all his case. But he has treated it with such a vast amount of original and collated observation, as fairly to have brought the subject within the bounds of rational scientific research.[68]

Interestingly, the committee making the award chose not to mention the *Origin of Species* explicitly, which caused Darwin's supporters some dissatisfaction. Therefore, at the anniversary dinner following the November meeting, it fell to Charles Lyell to make explicit what the committee did not: that Darwin was to be valued for his "capacity as a thinker and as a philosophical writer" that "would enable him to be author of such a book as the *Origin of Species*." In Lyell's words, "It is with that work that the public and more especially the rising generation will naturally recollect the honour which we are now doing to my friend."[69] In time, scientists concurred with Lyell's assessment. Darwin's theory of evolution through natural selection is now regarded as a cornerstone of modern biology.

Had Darwin, like Lincoln, died in the 1860s, his legacy in science would have been secure. The theory of evolution through natural selection had already been taken up by numerous investigators as a plausible explanation to account for the origin of species. However, the progress of scientific research benefited greatly from Darwin's good fortune of a longer life. One of his accomplishments during the 1860s and early 1870s was to fill in and complete the work begun in the *Origin*, which was, after all, only an abstract of what his longer book was to have been. He published what was to have been the first two chapters of his longer book under the title *The Variation of Animals and Plants under Domestication*.[70] His next book, *Descent of Man*, addressed the evolution of human beings, a controversial subject he had only touched on in the *Origin*.[71] As part of his evolutionary studies and to further illustrate a commonality between man and animals, Darwin then turned to the subject of emotion. *The Expression of the Emotions in Man and Animals* was published in 1872.[72] As if these efforts were not enough, Darwin devoted considerable energy after 1860 to the experimental study of plants, a subject on which he published several books.[73] For his last book, published in 1881, he returned to the subject of geology and ended his career with another experimental study, this one on the role of earthworms in creating soil.[74] Overall, Charles Darwin was a fine observer, a daring theorist, and a solid experimentalist whose efforts established modern biology on a new footing. He had taken up and answered the question of evolution.

NOTES

[1] For an excellent short introduction to Linnaeus's work, as well as to the naturalist tradition generally, see Paul Farber, *On Finding Order in Nature: The Naturalist Tradition from Linnaeus to E. O. Wilson* (Baltimore: Johns Hopkins University Press, 2000). On Linnaeus see also www.ucmp.berkeley.edu/history/linnaeus.html.

[2] Sandra Rebok, "Two Exponents of the Enlightenment: Transatlantic Communication by Thomas Jefferson and Alexander von Humboldt," *Southern Quarterly* 43, no. 4 (2006): 126–52.

[3] After their wives died, Jefferson and Erasmus Darwin each formed a liaison with a young unmarried woman who bore them children. Jefferson's mistress was Sally Hemings, a slave. On this relationship, see Annette Gordon-Reed, *The Hemingses of Monticello: An American Family* (New York: W. W. Norton, 2008). Darwin's mistress was Mary Parker, a servant in his household. Children of such unions were not treated by society as legal heirs. Darwin did, however, help his illegitimate daughters establish a boarding school in 1794, and he wrote a book on the subject of education for women: *A Plan for the Conduct of Female Education, in Boarding Schools* (1797).

[4] On the links among Small, Jefferson, and Erasmus Darwin, see Desmond King-Hele, *Erasmus Darwin: A Life of Unequalled Achievement* (London: DLM, 1999), 60. Small was assisted on his return to Britain by a letter of introduction from Benjamin Franklin.

[5] On the publishing history of Jefferson's book, see David Waldstreicher, ed., *Notes on the State of Virginia by Thomas Jefferson with Related Documents* (Boston: Bedford/

St. Martin's, 2002), 18–20; and Lee Alan Dugatkin, *Mr. Jefferson and the Giant Moose* (Chicago: University of Chicago Press, 2009), 62–68.

[6]Thomas Jefferson, "A Memoir on the Discovery of Certain Bones of a Quadruped of the Clawed Kind in the Western Parts of Virginia," *Transactions of the American Philosophical Society* 4 (1799): 246–60 [read on March 10, 1797]. In the printed article Jefferson was less dismissive of the possibility of extinction than he had been when he presented the paper two years earlier, for by this time he had become aware of Cuvier's identification of *Megatherium*. See Martin Rudwick, *Bursting the Limits of Time: The Reconstruction of Geohistory in the Age of Revolution* (Chicago: University of Chicago Press, 2005), 374–75.

[7]On Indian tradition, see Adrienne Mayor, *Fossil Legends of the First Americans* (Princeton, N.J.: Princeton University Press, 2005). For a photograph of a fossil mastodon femur collected by Indian hunters on the Ohio River in 1739, and now on display in Paris, see p. 18. On the mastodon, also see Paul Semonin, *American Monster: How the Nation's First Prehistoric Creature Became a Symbol of National Identity* (New York: New York University Press, 2000).

[8]Silvio A. Bedini, "Thomas Jefferson and American Vertebrate Paleontology," *Virginia Division of Mineral Resources Publication* 61 (Charlottesville, Va.: Department of Mines, Minerals and Energy, 1985), 17; and Rudwick, *Bursting the Limits of Time*, 374–75.

[9]Georges Cuvier, "Notice Concerning the Skeleton of a Very Large Species of Quadruped, Hitherto Unknown, Found at Paraguay, and Deposited in the Cabinet of Natural History at Madrid," *Monthly Magazine* 2 (1796): 638. The accompanying plate of *Megatherium* was similar to that reproduced in Document 6. For a finely demarcated treatment of Cuvier's development of the issue of extinction, see Martin J. S. Rudwick, *Georges Cuvier, Fossil Bones, and Geological Catastrophes* (Chicago: University of Chicago Press, 1997).

[10]Advertisement to Erasmus Darwin's *The Botanic Garden* (1791), quoted in Maureen McNeil, "Darwin, Erasmus (1731–1802)," *Oxford Dictionary of National Biography*, Oxford University Press, Sept. 2004; online edition, May 2007, www.oxforddnb.com/view/article/7177, accessed 8 Sept. 2010.

[11][Erasmus Darwin], *The Families of Plants, with Their Natural Characters . . . Translated from . . . the* Genera Plantarum . . . *of the Elder Linneus* (Lichfield, U.K., 1787). On Erasmus Darwin's botany, see Janet Browne, "Botany for Gentlemen: Erasmus Darwin and *The Loves of the Plants*," *Isis* 80 (1989), 593–621.

[12]Introduction to *The Families of Plants* (1787), xix–xx.

[13]Erasmus Darwin, *Temple of Nature*, 38 of "Additional Notes." *Ens entium* is Latin for "being of beings."

[14]Erasmus Darwin to Richard Lovell Edgeworth, March 15, 1795, quoted in King-Hele, *Erasmus Darwin*, 295.

[15]Thomas Robert Malthus, *An Essay on the Principle of Population, as It Affects the Future Improvement of Society, with Remarks on the Speculations of Mr. Godwin, M. Condorcet, and Other Writers* (London: J. Johnson, 1798), 1–2. For background, see Gail Bederman, "Sex, Scandal, Satire, and Population in 1798: Revisiting Malthus's First Essay," *Journal of British Studies* 47 (2008): 768–95.

[16]William Paley, *Natural Theology, or Evidence of the Existence and Attributes of the Deity, Collected from the Appearances of Nature*, edited with an introduction and notes by Matthew D. Eddy and David Knight (Oxford, U.K.: Oxford University Press, 2006).

[17]See the editors' notes in Paley, *Natural Theology*, 325.

[18]Gordon L. Herries Davies, *Whatever Is under the Earth: The Geological Society of London: 1807 to 2007* (London: Geological Society, 2007).

[19]Malthus, *An Essay on the Principle of Population*, 210.

[20]Dorinda Outram, "New Species in Natural History," in *Cultures of Natural History*, ed. N. Jardine, J. A. Secord, and E. C. Spary (Cambridge, U.K.: Cambridge University Press, 1996), 251.

[21]Frederick Burkhardt Jr., "The Leopard in the Garden: Life in Close Quarters at the Muséum d'Histoire Naturelle," *Isis* 98 (2007): 675–94.

[22]Georges Cuvier, *Recherches sur les ossemens fossiles de quadrupèdes*, 4 vols. (Paris: Deterville, 1812).

[23]Richard W. Burkhardt Jr., "The Inspiration of Lamarck's Belief in Evolution," *Journal of the History of Biology* 5 (1972): 421.

[24]An English translation is available as J. B. Lamarck, *Zoological Philosophy: An Exposition with Regard to the Natural History of Animals,* trans. Hugh Elliot and with introductory essays by David L. Hull and Richard W. Burkhardt Jr. (Chicago: University of Chicago Press, 1984). The volume also contains a translation of Lamarck's lecture of 1800.

[25]Peter Bowler, "The Changing Meaning of Evolution," *Journal of the History of Ideas* 36 (1975): 95–114; and Rudwick, *Bursting the Limits of Time*, 245. It should also be noted that where Lamarck chose to be precise, he could be. He certainly believed in spontaneous generation and used the words *génération directe* or *spontanée.*

[26]See www.Lamarck.net, edited by Pietro Corsi, for the database concerning the 973 pupils who attended Lamarck's lectures from 1795–1823. (Accessed February 28, 2008.) According to Corsi (personal communication, January 21, 2006), while Erasmus Darwin died too soon to have heard of Lamarck's transformist views, there is no evidence either way that Lamarck had read Erasmus Darwin. Corsi stressed, however, that there were other European thinkers whose views paralleled those of Erasmus Darwin.

[27]Pietro Corsi, "Before Darwin: Transformist Concepts in European Natural History," *Journal of the History of Biology* 38 (2005): 67–83.

[28]Quoted in Hervé Le Guyader, *Geoffroy Saint-Hilaire: A Visionary Naturalist*, trans. Marjorie Grene (Chicago: University of Chicago Press, 2004), p. 21. The standard work on this subject is Toby A. Appel, *The Cuvier-Geoffroy Debate: French Biology in the Decades before Darwin* (Oxford, U.K.: Oxford University Press, 1987). The debate between Cuvier and Geoffroy came to a head in 1830.

[29]Nora Barlow, ed., *The Autobiography of Charles Darwin, 1809–1882* (1958; reprint, New York: W. W. Norton and Co., 1969). Also see the entry on Robert Grant by Adrian Desmond in the *Oxford Dictionary of National Biography.*

[30]Charles Lyell, *Principles of Geology, Being an Attempt to Explain the Former Changes of the Earth's Surface, by Reference to Causes Now in Operation*, 3 vols. (London: John Murray, 1830–1833).

[31]C. Darwin to Leonard Horner, August 29, [1844], in *The Correspondence of Charles Darwin*, eds. Frederick Burkhardt et al., 16+ vols. (Cambridge, U.K.: Cambridge University Press, 1985–), 3:55. Also available online at www.darwinproject.ac.uk.

[32]Charles Lyell to Gideon Mantell, March 2, 1827, in *Life, Letters and Journals of Sir Charles Lyell, Bart.*, ed. Katharine Lyell (London: John Murray, 1881), 1:168.

[33]David Kohn, Gina Murrell, John Parker, and Mark Whitehorn, "What Henslow Taught Darwin," *Nature* 436 (2005): 643–45.

[34]J. S. Henslow to C. Darwin, August 24, 1831, in *The Correspondence of Charles Darwin*, 1:129.

[35]The list of zoological specimens from the voyage has been published in Richard Darwin Keynes, ed., *Charles Darwin's Zoology Notes and Specimen Lists from H.M.S. Beagle* (Cambridge, U.K.: Cambridge University Press, 2000).

[36]Robert FitzRoy, ed., *Narrative of the Surveying Voyages of His Majesty's Ships Adventure and Beagle*, vol. 3: *Journal and Remarks, 1832–1836*, by Charles Darwin (London: Henry Colburn, 1839), 454–55.

[37]Sandra Herbert, ed., *The Red Notebook of Charles Darwin* (Ithaca, N.Y.: Cornell University Press, 1980), 66 [RN:130]. For fuller treatments of Darwin's adoption of transmutation, see my introduction to the *Red Notebook*, pp. 5–29; and Frank J. Sulloway, "Darwin's Conversion: The *Beagle* Voyage and Its Aftermath," *Journal of the History of Biology*, 15:325–96, including p. 350 for type specimens of the mockingbird. For an excellent representation of the differences among present-day Galápagos mockingbirds,

see K. Thalia Grant and Gregory B. Estes, *Darwin in Galápagos* (Princeton, N.J.: Princeton University Press, 2009), plate 12. To view Darwin's specimens of Galápagos mockingbirds, see www.nhm.ac.uk/nature-online/evolution/how-did-evol-theory-develop/galapagos-mockingbirds/index.html.

[38]Erasmus Darwin, *Zoonomia; or, the Laws of Organic Life,* 2 vols. (London: J. Johnson, 1794, 1796); and Paul H. Barrett, Peter J. Gautrey, Sandra Herbert, David Kohn, and Sydney Smith, eds., *Charles Darwin's Notebooks, 1836–1844* (Ithaca, N.Y.: Cornell University Press, 1987), 170 [B:1].

[39]*Charles Darwin's Notebooks*, 172 [B:7].

[40]*Charles Darwin's Notebooks*, 228 [B:231–32].

[41]*Charles Darwin's Notebooks*, 195 [B:101].

[42]*Charles Darwin's Notebooks*, 291 [C:166].

[43]Charles Darwin to J. D. Hooker, July 12, 1870, in Frederick Burkhardt, Samantha Evans, and Alison Pearn, eds., *Evolution: Selected Letters of Charles Darwin, 1860–1870* (Cambridge, U.K.: Cambridge University Press, 2008), 247.

[44]*Charles Darwin's Notebooks*, 375–76 [D:135].

[45]*Charles Darwin's Notebooks*, 639 [Notes on Macculloch:28^{v}].

[46]The questionnaire is reprinted in *The Correspondence of Charles Darwin*, 2:446–49. Also see "Questions & Experiments" in *Charles Darwin's Notebooks*, 487–516.

[47]Louis Agassiz, "Upon Glaciers, Moraines and Erratic Blocks; Being the Address Delivered at the Opening of the Helvetic Natural History Society at Neuchatel [*sic*], on 24th July 1837," *Edinburgh New Philosophical Journal* 24 (1838): 364–83. For Darwin's treatment of Agassiz's views on glaciers, see Charles Darwin, *Journal of Researches* (1839), 617–26.

[48]Charles Darwin, *Geological Observations on South America. Being the Third Part of the Geology of the Voyage of the "Beagle"* (London: Smith, Elder, and Co., 1846).

[49]The essays are contained in Charles Darwin and Alfred Russel Wallace, *Evolution by Natural Selection* (Cambridge, U.K.: Cambridge University Press, 1958).

[50]Charles Darwin to Emma Darwin, July 5, 1844, in *The Correspondence of Charles Darwin*, 4:43–45.

[51]James A. Secord, *Victorian Sensation: The Extraordinary Publication, Reception, and Secret Authorship of "Vestiges of the Natural History of Creation"* (Chicago: University of Chicago Press, 2000), 131, 2.

[52]Secord, *Victorian Sensation*, 1.

[53][Robert Chambers], *Vestiges of the Natural History of Creation* (London: John Churchill, 1844), 191.

[54]Alfred Russel Wallace, *A Narrative of Travels on the Amazon and Rio Negro, with an Account of the Native Tribes, and Observations on the Climate, Geology, and Natural History of the Amazon Valley* (London: Reeve and Co., 1853); and Alfred Russel Wallace, *Palm Trees of the Amazon and Their Uses* (London: J. Van Voorst, 1853).

[55]Alfred Russel Wallace, *The Malay Archipelago: The Land of the Orang-utan and the Bird of Paradise: A Narrative of Travel with Studies of Man and Nature*, 2 vols. (London: Macmillan, 1869).

[56]Robert Stauffer, ed., *Charles Darwin's Natural Selection: Being the Second Part of His Big Species Book Written from 1856 to 1858* (Cambridge, U.K.: Cambridge University Press, 1975).

[57]The publication appeared as Charles Darwin and Alfred Wallace, "On the Tendency of Species to Form Varieties and on the Perpetuation of Varieties and Species by Natural Means of Selection," *Journal of the Linnean Society of London (Zoology)* 3 (1858): 45–62. Images of the text can be viewed at www.darwinproject.ac.uk.

[58]See Thomas F. Glick, ed., *The Comparative Reception of Darwinism* (Chicago: University of Chicago Press, 1988); and Thomas F. Glick, Miguel Angel Puig-Samper, and Rosaura Ruiz, *The Reception of Darwinism in the Iberian World: Spain, Spanish America, and Brazil* (Dordrecht: Kluwer Academic, 2001).

[59]Michèle Kohler and Chris Kohler, "The *Origin of Species* as a Book," in Michael Ruse and Robert J. Richards, *The Cambridge Companion to the "Origin of Species"* (Cambridge, U.K.: Cambridge University Press, 2009), 340–42.

[60]Joy Harvey, *Almost a Man of Genius: Clémence Royer, Feminism, and Nineteenth-Century Science* (New Brunswick, N.J.: Rutgers University Press, 1997).

[61]See the classic analysis in Alvar Ellegård, *Darwin and the General Reader: The Reception of Darwin's Theory of Evolution in the British Periodical Press, 1859–1872,* with a new foreword by David L. Hull (Chicago: University of Chicago Press, 1990).

[62]On the 1860 meeting of the British Association for the Advancement of Science, see J. V. Jensen, "Return to the Wilberforce-Huxley Debate," *British Journal for the History of Science* 21 (1988): 161–79.

[63]See Edward Larson, *Summer for the God: The Scopes Trial and America's Continuing Debate over Science and Religion* (New York: Basic Books, 1997); and Jeffrey P. Moran, *The Scopes Trial: A Brief History with Documents* (Boston: Bedford/St. Martin's, 2002).

[64]Asa Gray, "Review of Darwin's Theory on the Origin of Species by Means of Natural Selection," *American Journal of Science and Arts*, 2nd ser., 29 (1860): 153–84.

[65]In 1835 Henslow had a collection of Darwin's letters printed privately. Henry's copy of this tract is presently held by the Smithsonian Institution Libraries.

[66]Joseph Henry, Document 134 (March 16, 1862), in Marc Rothenberg, ed., *The Papers of Joseph Henry* (Smithsonian Institution/Science History Publications, 2004), 10:244.

[67]On Lincoln's career, see James M. McPherson, "Abraham Lincoln," in *American National Biography*, edited by John A. Garraty and Mark C. Carnes (Oxford, U.K.: Oxford University Press, 1999), 13:662–73.

[68]Hugh Falconer to William Sharpey, October 25, 1864, in *The Correspondence of Charles Darwin*, 12:377–78.

[69]Charles Lyell, November 30, 1864, quoted in *The Correspondence of Charles Darwin*, 12:520.

[70]Charles Darwin, *The Variation of Animals and Plants under Domestication*, 2 vols. (London: John Murray, 1868).

[71]Charles Darwin, *The Descent of Man, and Selection in Relation to Sex*, 2 vols. (London: John Murray, 1870, 1871).

[72]Charles Darwin, *The Expression of the Emotions in Man and Animals* (London: John Murray, 1872).

[73]See R. B. Freeman, *The Works of Charles Darwin: An Annotated Bibliographical Handlist*, 2nd ed. (London: Dawson, 1977). This book can be accessed on www.darwin-online.org.uk.

[74]Charles Darwin, *The Formation of Vegetable Mould, through the Action of Worms, with Observations on Their Habits* (London: John Murray, 1881).

PART TWO

The Documents

1
The Question of Evolution Arises

1

CARL LINNAEUS

Genera Plantarum: The Families of Plants

1787

Swedish physician and natural historian Carl von Linné (1707–1778), or Linnaeus as he is usually called, advocated the idea that every species should have only one scientific name and that the name should be in Latin. One of the people who promulgated Linnaeus's idea in the English-speaking world was Erasmus Darwin, Charles Darwin's grandfather. He supervised a translation of Linnaeus's Genera Plantarum, which was published under the auspices of "a Botanical Society at Lichfield." As Erasmus Darwin wrote to Benjamin Franklin, his intention in publishing the book was to "propagate the knowledge of Botany," though he doubted that the work would be worth reprinting in America since he imagined that only about twenty sets of the book might be sold.[1] *One of the features of Linnaeus's botanical work was that it emphasized the sexuality of plants, a point on which Darwin elaborated in a number of his writings. The photographs shown here are the inside front cover with Darwin's handwritten inscription and the book's title page. This copy is housed in the rare book room at the Library of Congress. Darwin's gift suggests the close ties between his Lunar Society circle and the scientifically minded Franklin.*

[1] Erasmus Darwin to Benjamin Franklin, May 29, 1787, quoted in Desmond King-Hele, *Erasmus Darwin: A Life of Unparalleled Achievement* (London: Giles de la Mare, 1999), 219.

Carl Linnaeus, *Genera Plantarum: The Families of Plants, with Their Natural Characters, according to the Number, Figure, Situation, and Proportion of All of the Parts of Fructification*, 2 vols. (Lichfield, U.K.: J. Johnson, 1787).

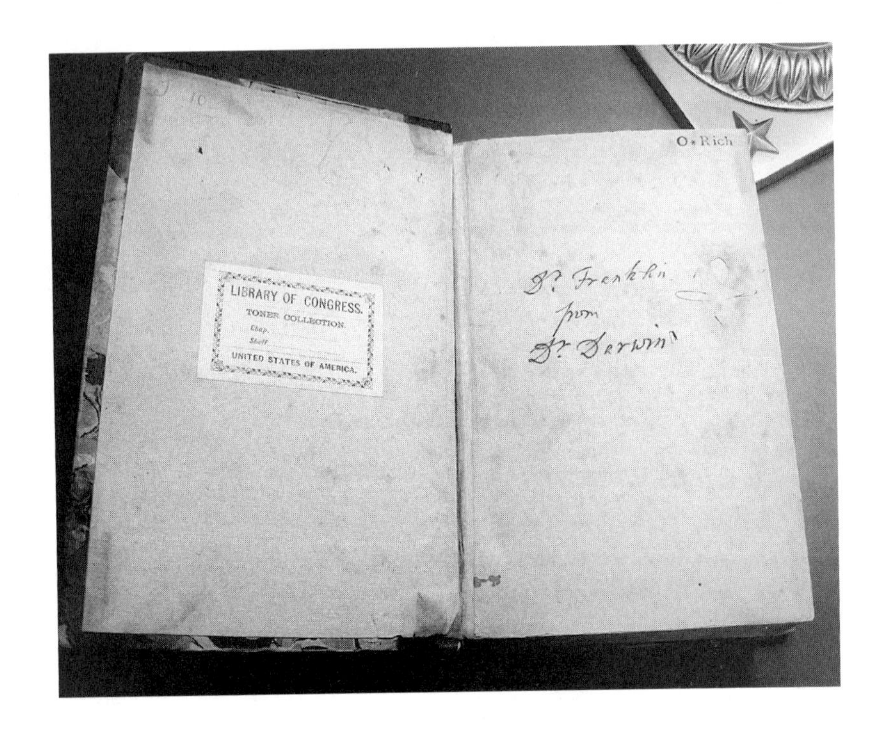
O. Rich
LIBRARY OF CONGRESS.
TONER COLLECTION.
Chap.
Shelf
UNITED STATES OF AMERICA.

THE

FAMILIES

OF

PLANTS,

WITH THEIR NATURAL CHARACTERS,

ACCORDING TO THE

NUMBER, FIGURE, SITUATION, AND *PROPORTION*

OF ALL THE PARTS OF FRUCTIFICATION.

TRANSLATED FROM THE LAST EDITION,

(AS PUBLISHED BY DR. REICHARD)

OF THE

GENERA PLANTARUM,

AND OF THE

MANTISSÆ PLANTARUM

OF THE ELDER LINNEUS;

AND FROM THE

SUPPLEMENTUM PLANTARUM

OF THE YOUNGER LINNEUS,

WITH ALL THE NEW

FAMILIES OF PLANTS,

FROM

THUNBERG AND *L'HERITIER.*

TO WHICH IS PREFIX'D AN ACCENTED CATALOGUE OF THE NAMES OF PLANTS, WITH THE ADJECTIVES APPLY'D TO THEM, AND OTHER BOTANIC TERMS, FOR THE PURPOSE OF TEACHING THEIR RIGHT PRONUNCIATION.

VOL. I.

BY A BOTANICAL SOCIETY AT LICHFIELD.

LICHFIELD: PRINTED BY *JOHN JACKSON.*
SOLD BY J. JOHNSON, ST. PAUL'S CHURCH-YARD, LONDON.
T. BYRNE, DUBLIN. AND J. BALFOUR, EDINBURGH.

MDCCLXXXVII.

2

ALEXANDER VON HUMBOLDT

Personal Narrative of Travels to the Equinoctial Regions of the New Continent, during the Years 1799–1804

1814

Alexander von Humboldt (1769–1859) was a great traveler and geographer. From 1799 through 1804 he traveled in the New World, spending the most time in northern South America, Mexico, and the Caribbean but also venturing up to the United States. He described his goal as a scientific traveler as follows: "I had in view a double purpose in the travels, of which I now publish the historical narrative. I wished to make known the countries I had visited; and to collect such facts as are fitted to elucidate a science, of which we have possessed scarcely the outline, and which has been vaguely denominated natural history of the world, theory of the Earth, *or physical* geography.*" It was not only scientific interest that drove Humboldt to travel: He also wanted to experience for himself distant worlds, fulfilling a childhood longing. His love of travel comes out strongly in the excerpt reprinted here. In it he describes his excitement on crossing the equator, entering the Southern Hemisphere, and seeing in the night sky the Magellanic Clouds and the Southern Cross, neither of which is visible from northern latitudes. Humboldt's vivid depiction of the beauty of the skies of the Southern Hemisphere inspired in some readers a desire to see the sight for themselves.*

From the time we entered the torrid zone, we were never wearied with admiring, every night, the beauty of the southern sky, which, as we advanced towards the south, opened new constellations to our view. We feel an indescribable sensation, when, on approaching the equator, and

---

From Alexander von Humboldt, *Personal Narrative of Travels to the Equinoctial Regions of the New Continent, during the Years 1799–1804*, trans. Helen Maria Williams, 7 vols. (London: Longman, 1814–1829). The quotation in the headnote is from 1:iii. The excerpt reprinted here is from 2:18–20.

particularly on passing from one hemisphere to the other, we see those stars, which we have contemplated from our infancy, progressively sink, and finally disappear. Nothing awakens in the traveller a livelier remembrance of the immense distance by which he is separated from his country, than the aspect of an unknown firmament. The grouping of the stars of the first magnitude, some scattered nebulæ, rivalling in splendor the milky way, and tracks of space remarkable for their extreme blackness, give a particular physiognomy to the southern sky. This sight fills with admiration even those, who, uninstructed in the branches of accurate science, feel the same emotion of delight in the contemplation of the heavenly vault, as in the view of a beautiful landscape, or a majestic site. A traveller has no need of being a botanist, to recognize the torrid zone on the mere aspect of its vegetation; and without having acquired any notions of astronomy, without any acquaintance with the celestial charts of Flamstead and de la Caille,[1] he feels he is not in Europe, when he sees the immense constellation of the Ship, or the phosphorescent clouds of Magellan, arise on the horizon. The heaven, and the earth, every thing in the equinoctial regions, assumes an exotic character.

The lower regions of the air were loaded with vapors for some days. We saw distinctly for the first time the Cross of the south only in the night of the 4th and 5th of July, in the sixteenth degree of latitude; it was strongly inclined, and appeared from time to time between the clouds, the centre of which, furrowed by uncondensed lightnings, reflected a silver light. If a traveller may be permitted to speak of his personal emotions, I shall add, that in this night I saw one of the reveries of my earliest youth accomplished.

[1] John Flamsteed (1646–1719), English astronomer. Nicolas Louis de Lacaille (1713–1762), French astronomer.

3

ERASMUS DARWIN

The Temple of Nature; or, the Origin of Society

1803

This poem was Erasmus Darwin's last work, published a year after his death. It contains the most succinct statement of his evolutionary views: "Organic Life beneath the shoreless waves / Was born and nurs'd in Ocean's pearly caves." Like Darwin's earlier works, it influenced the Romantic poets, not so much in its poetic style, which was rather heavy, but in its emphasis on nature. As with all of this author's works, emphasis is on the progressive aspect of life. Darwin's original preface and prose footnotes that accompany his poem (edited here for consistency with the excerpt) form an important part of the text and convey what was then current scientific knowledge.

Preface

The Poem, which is here offered to the Public, does not pretend to instruct by deep researches of reasoning; its aim is simply to amuse by bringing distinctly to the imagination the beautiful and sublime images of the operations of Nature in the order, as the Author believes, in which the progressive course of time presented them. . . .

Canto I. Production of Life

V. "Organic Life beneath the shoreless waves [295]
Was born and nurs'd in Ocean's pearly caves;

Beneath the shoreless waves, l. 295. The earth was originally covered with water, as appears from some of its highest mountains, consisting of shells cemented together by a solution of part of them, as the limestone rocks of the Alps; Ferber's Travels. It must be therefore concluded, that animal life began beneath the sea.

From Erasmus Darwin, *The Temple of Nature; or, the Origin of Society: A Poem, with Philosophical Notes* (London: J. Johnson, 1803).

First forms minute, unseen by spheric glass,
Move on the mud, or pierce the watery mass;
These, as successive generations bloom,
New powers acquire, and larger limbs assume; 300
Whence countless groups of vegetation spring,
And breathing realms of fin, and feet, and wing.

"Thus the tall Oak, the giant of the wood,
Which bears Britannia's thunders on the flood;
The Whale, unmeasured monster of the main,
The lordly Lion, monarch of the plain,
The Eagle soaring in the realms of air,
Whose eye undazzled drinks the solar glare,
Imperious man, who rules the bestial crowd,
Of language, reason, and reflection proud, 310
With brow erect who scorns this earthy sod,
And styles himself the image of his God;
Arose from rudiments of form and sense,
An embryon point, or microscopic ens![1]

. . .

"Next when imprison'd fires in central caves
Burst the firm earth, and drank the headlong waves;
And, as new airs with dread explosion swell,
Form'd lava-isles, and continents of shell;

Nor is this unanalogous to what still occurs, as all quadrupeds and mankind in their embryon state are aquatic animals; and thus may be said to resemble gnats and frogs. The fetus in the uterus has an organ called the placenta, the fine extremities of the vessels of which permeate the arteries of the uterus, and the blood of the fetus becomes thus oxygenated from the passing stream of the maternal arterial blood; exactly as is done by the gills of fish from the stream of water, which they occasion to pass through them.

But the chicken in the egg possesses a kind of aerial respiration, since the extremities of its placental vessels terminate on a membranous bag, which contains air, at the broad end of the egg; and in this the chick in the egg differs from the fetus in the womb, as there is in the egg no circulating maternal blood for the insertion of the extremities of its respiratory vessels, and in this also I suspect that the eggs of birds differ from the spawn of fish; which latter is immersed in water, and which has probably the extremities of its respiratory organ inserted into the soft membrane which covers it, and is in contact with the water.

An embryon point, l. 314. The arguments showing that all vegetables and animals arose from such a small beginning, as a living point or living fibre, are detailed in Zoonomia, Sect. XXXIX. 4. 8. on Generation.

[1] *ens*: Latin: being.

Pil'd rocks on rocks, on mountains mountains raised,
And high in heaven the first volcanoes blazed;
In countless swarms an insect-myriad moves
From sea-fan gardens, and from coral groves;
Leaves the cold caverns of the deep, and creeps
On shelving shores, or climbs on rocky steeps. 330
As in dry air the sea-born stranger roves,
Each muscle quickens, and each sense improves;
Cold gills aquatic form respiring lungs,
And sounds aerial flow from slimy tongues.

"So Trapa rooted in pellucid tides,
In countless threads her breathing leaves divides,

An insect-myriad moves, l. 327. After islands or continents were raised above the primeval ocean, great numbers of the most simple animals would attempt to seek food at the edges or shores of the new land, and might thence gradually become amphibious; as is now seen in the frog, who changes from an aquatic animal to an amphibious one; and in the gnat, which changes from a natant to a volant state.

At the same time new microscopic animalcules would immediately commence wherever there was warmth and moisture, and some organic matter, that might induce putridity. Those situated on dry land, and immersed in dry air, may gradually acquire new powers to preserve their existence; and by innumerable successive reproductions for some thousands, or perhaps millions of ages, may at length have produced many of the vegetable and animal inhabitants which now people the earth.

As innumerable shell-fish must have existed a long time beneath the ocean, before the calcareous mountains were produced and elevated; it is also probable, that many of the insect tribes, or less complicate animals, existed long before the quadrupeds or more complicate ones, which in some measure accords with the theory of Linneus in respect to the vegetable world; who thinks, that all the plants now extant arose from the conjunction and reproduction of about sixty different vegetables, from which he constitutes his natural orders.

As the blood of animals in the air becomes more oxygenated in their lungs, than that of animals in water by their gills; it becomes of a more scarlet colour, and from its greater stimulus the sensorium seems to produce quicker motions and finer sensations; and as water is a much better vehicle for vibrations or sounds than air, the fish, even when dying in pain, are mute in the atmosphere, though it is probable that in the water they may utter sounds to be heard at a considerable distance. See on this subject, Botanic Garden, Vol. I. Canto IV. l. 176, Note.

So Trapa rooted, l. 335. The lower leaves of this plant grow under water, and are divided into minute capillary ramifications; while the upper leaves are broad and round, and have air bladders in their footstalks to support them above the surface of the water. As the aerial leaves of vegetables do the office of lungs, by exposing a large surface of vessels with their contained fluids to the influence of the air; so these aquatic leaves answer a similar purpose like the gills of fish, and perhaps gain from water a similar material. As the material thus necessary to life seems to be more easily acquired from air than from water, the subaquatic leaves of this plant and of sisymbrium, oenanthe, ranunculus aquatilis, water crow foot, and some others, are cut into fine divisions to increase the surface, whilst those above water are undivided; see Botanic Garden, Vol. II. Canto IV. l. 204, Note.

Waves her bright tresses in the watery mass,
And drinks with gelid[2] gills the vital gas;
Then broader leaves in shadowy files advance,
Spread o'er the crystal flood their green expanse; 340
And, as in air the adherent dew exhales,
Court the warm sun, and breathe ethereal gales.

"So still the Tadpole cleaves the watery vale
With balanc'd fins, and undulating tail;
New lungs and limbs proclaim his second birth,
Breathe the dry air, and bound upon the earth.
So from deep lakes the dread Musquito springs,
Drinks the soft breeze, and dries his tender wings,
In twinkling squadrons cuts his airy way,
Dips his red trunk in blood, and man his prey. 350

"So still the Diodons,[3] amphibious tribe,
With two-fold lungs the sea or air imbibe;
Allied to fish, the lizard cleaves the flood
With one-cell'd heart, and dark frigescent[4] blood;
Half-reasoning Beavers long-unbreathing dart
Through Erie's waves with perforated heart;
With gills and lungs respiring Lampreys steer,
Kiss the rude rocks, and suck till they adhere;

Few of the water plants of this country are used for economical purposes, but the ranunculus fluviatilis may be worth cultivation; as on the borders of the river Avon, near Ringwood, the cottagers cut this plant every morning in boats, almost all the year round, to feed their cows, which appear in good condition, and give a due quantity of milk; see a paper from Dr. Pultney in the Transactions of the Linnean Society, Vol. V.

So still the Tadpole, l. 343 The transformation of the tadpole from an aquatic animal into an aerial one is abundantly curious. When first it is hatched from the spawn by the warmth of the season, it resembles a fish; it afterwards puts forth legs, and resembles a lizard; and finally losing its tail, and acquiring lungs instead of gills, becomes an aerial quadruped.

The rana temporaria of Linneus lives in the water in spring, and on the land in summer, and catches flies. Of the rana paradoxa the larva or tadpole is as large as the frog, and dwells in Surinam, whence the mistake of Merian and of Seba, who call it a frog fish. The esculent frog is green, with three yellow lines from the mouth to the anus; the back transversely gibbous, the hinder feet palmated; its more frequent croaking in the evenings is said to foretell rain. Linnei Syst. Nat. Art. rana.

Linneus asserts in his introduction to the class Amphibia, that frogs are so nearly allied to lizards, lizards to serpents, and serpents to fish, that the boundaries of these orders can scarcely be ascertained.

[2] *gelid*: frosty.
[3] *Diodons*: genus including porcupine puffer fish.
[4] *frigescent*: cooling.

The lazy Remora's inhaling lips,
Hung on the keel, retard the struggling ships; 360
With gills pulmonic breathes the enormous Whale,
And spouts aquatic columns to the gale;
Sports on the shining wave at noontide hours,
And shifting rainbows crest the rising showers.

"So erst, ere rose the science to record
In letter'd syllables the volant word;
Whence chemic arts, disclosed in pictured lines,
Liv'd to mankind by hieroglyphic signs;
And clustering stars, pourtray'd on mimic spheres,
Assumed the forms of lions, bulls, and bears; 370
—So erst, as Egypt's rude designs explain,
Rose young Dione from the shoreless main;
Type of organic Nature! source of bliss!
Emerging Beauty from the vast abyss!
Sublime on Chaos borne, the Goddess stood,
And smiled enchantment on the troubled flood;
The warring elements to peace restored,
And young Reflection wondered and adored."

Now paused the Nymph,—The Muse responsive cries,
Sweet admiration sparkling in her eyes, 380
"Drawn by your pencil, by your hand unfurl'd,
Bright shines the tablet of the dawning world;
Amazed the Sea's prolific depths I view,
And Venus rising from the waves in You!

At noontide hours, l. 363. The rainbows in our latitude are only seen in the mornings or evenings, when the sun is not much more than forty-two degrees high. In the more northern latitudes, where the meridian sun is not more than forty-two degrees high, they are also visible at noon.

Rose young Dione, l. 372. The hieroglyphic figure of Venus rising from the sea supported on a shell by two tritons, as well as that of Hercules armed with a club, appear to be remains of the most remote antiquity. As the former is devoid of grace, and of the pictorial art of design, as one half of the group exactly resembles the other; and as that of Hercules is armed with a club, which was the first weapon.

The Venus seems to have represented the beauty of organic Nature rising from the sea, and afterwards became simply an emblem of ideal beauty; while the figure of Adonis was probably designed to represent the more abstracted idea of life or animation. Some of these hieroglyphic designs seem to evince the profound investigations in science of the Egyptian philosophers, and to have outlived all written language; and still constitute the symbols, by which painters and poets give form and animation to abstracted ideas, as to those of strength and beauty in the above instances.

"Still Nature's births enclosed in egg or seed
From the tall forest to the lowly weed,
Her beaux and beauties, butterflies and worms,
Rise from aquatic to aerial forms.
Thus in the womb the nascent infant laves
Its natant form in the circumfluent waves; 390
With perforated heart unbreathing swims,
Awakes and stretches all its recent limbs;
With gills placental seeks the arterial flood,
And drinks pure ether from its Mother's blood.
Erewhile the landed Stranger bursts his way,
From the warm wave emerging into day;
Feels the chill blast, and piercing light, and tries
His tender lungs, and rolls his dazzled eyes;
Gives to the passing gale his curling hair,
And steps a dry inhabitant of air. 400

"Creative Nile, as taught in ancient song,
So charm'd to life his animated throng;
O'er his wide realms the slow-subsiding flood
Left the rich treasures of organic mud;
While with quick growth young Vegetation yields
Her blushing orchards, and her waving fields;
Pomona's hand replenish'd Plenty's horn,
And Ceres laugh'd amid her seas of corn.—
Bird, beast, and reptile, spring from sudden birth,
Raise their new forms, half-animal, half-earth; 410
The roaring lion shakes his tawny mane,
His struggling limbs still rooted in the plain;

Awakes and stretches, l. 392. During the first six months of gestation, the embryon probably sleeps, as it seems to have no use for voluntary power; it then seems to awake, and to stretch its limbs, and change its posture in some degree, which is termed quickening.

With gills placental, l. 393. The placenta adheres to any side of the uterus in natural gestation, or of any other cavity in extrauterine gestation; the extremities of its arteries and veins probably permeate the arteries of the mother, and absorb from thence through their fine coats the oxygen of the mother's blood; hence when the placenta is withdrawn, the side of the uterus, where it adhered, bleeds; but not the extremities of its own vessels.

His dazzled eyes, l. 398. Though the membrana pupillaris described by modern anatomists guards the tender retina from too much light; the young infant nevertheless seems to feel the presence of it by its frequently moving its eyes, before it can distinguish common objects.

With flapping wings assurgent eagles toil
To rend their talons from the adhesive soil;
The impatient serpent lifts his crested head,
And drags his train unfinish'd from the bed.—
As Warmth and Moisture blend their magic spells,
And brood with mingling wings the slimy dells;
Contractile earths in sentient forms arrange,
And Life triumphant stays their chemic change." 420

Then hand in hand along the waving glades
The virgin Sisters pass beneath the shades;
Ascend the winding steps with pausing march,
And seek the Portico's susurrant[5] arch;
Whose sculptur'd architrave on columns borne
Drinks the first blushes of the rising morn,
Whose fretted roof an ample shield displays,
And guards the Beauties from meridian rays.
While on light step enamour'd Zephyr springs,
And fans their glowing features with his wings, 430
Imbibes the fragrance of the vernal flowers,
And speeds with kisses sweet the dancing Hours. . . .

First her sweet voice in plaintive accents chains
The Muse's ear with fascinating strains;
Reverts awhile to elemental strife,

As Warmth and Moisture, l. 417.

In eodem corpore sæpe
Altera pars vivit; rudis est pars altera tellus.
Quippe ubi temperiem sumpsêre humorque calorque,
Concipiunt; & ab his oriuntur, cuncta duobus.

Ovid. Met. 1. l. 430.

Translation:

"And oft-times in the same body one part is alive and the other still nothing but raw earth. For when moisture and heat unite, life is conceived, and from these two sources all living things spring." (Ovid, *Metamorphoses*, translation by Frank Justus Miller, 3rd ed. rev by G. P. Goold, 2 vols. [Cambridge, Mass.: Harvard University Press, 1984] 1, 33)—Ed.]

This story from Ovid of the production of animals from the mud of the Nile seems to be of Egyptian origin, and is probably a poetical account of the opinions of the magi or priests of that country; showing that the simplest animations were spontaneously produced like chemical combinations, but were distinguished from the latter by their perpetual improvement by the power of reproduction, first by solitary, and then by sexual generation; whereas the products of natural chemistry are only enlarged by accretion, or purified by filtration.

[5] *susurrant*: whispering or murmuring.

The change of form, and brevity of life;
Then tells how potent Love with torch sublime
Relights the glimmering lamp, and conquers Time.

4

JOSIAH WEDGWOOD

"Am I Not a Man and a Brother?"

1787

An American Version, 1837

Abolitionists believed in the natural right of man to liberty. Their belief in this cause illustrates how closely attitudes toward nature and the "natural" were aligned with attitudes toward political rights. In 1787, the pottery firm of Josiah Wedgwood (Charles Darwin's maternal grandfather) issued this medallion to support the work of the abolitionist movement. Wedgwood sent several copies of the medallion to Benjamin Franklin for his use in campaigning against slavery in the United States. The image of the kneeling slave soon became a familiar symbol of abolitionist sympathies on both sides of the Atlantic.

The version of the image shown here was featured on a large political broadside issued at the Anti-Slavery Office on Nassau Street in New York City in 1837, as an unsigned woodcut printed on wove paper. It included a poem by John Greenleaf Whittier (1807–1892) entitled "Our Countrymen in Chains." The text on the broadside also declared, "England has 800,000 Slaves, and she has made them FREE. America has 2,250,000 Slaves, and she HOLDS THEM FAST!!!" This statement refers to the fact that Britain freed the slaves in its colonies in the mid-1830s while slavery still existed in the United States.

The Library of Congress (image and text available online under "Wedgwood" in the Prints and Photographs Online Catalog). For comparison, see the original Wedgwood image on the British Museum Web site (www.britishmuseum.org under the heading "Anti-slavery medallion by Josiah Wedgwood"). I thank Gloria Oden for directing me to the American versions of the original image.

AM I NOT A MAN AND A BROTHER?

5

THOMAS JEFFERSON

Notes on the State of Virginia

1787

Thomas Jefferson served, along with Benjamin Franklin, as a prominent mediator between the high culture of Enlightenment Europe and the newly formed United States of America. In Notes on the State of Virginia, *his only published book, Jefferson described in encyclopedic fashion the natural productions and the society of his native land. In the following passages excerpted from that book Jefferson describes the mammoth—his description was based partly on Indian accounts—and then suggests that the animal may have retired to the northern and western parts of America, where he hoped that future explorers might find it. Jefferson did not at this time believe that species became extinct.*

Our quadrupeds have been mostly described by Linnæus and Mons. de Buffon. Of these the mammoth, or big buffalo, as called by the Indians, must certainly have been the largest. Their tradition is, that he was carnivorous, and still exists in the Northern parts of America. A delegation of warriors from the Delaware tribe having visited the governor of Virginia, during the present revolution, on matters of business, after these had been discussed and settled in council, the governor asked them some questions relative to their country, and, among others, what they knew or had heard of the animal whose bones were found at the Saltlicks, on the Ohio. Their chief speaker immediately put himself into an attitude of oratory, and with a pomp suited to what he conceived the elevation of his subject, informed him that it was a tradition handed down from their fathers, "That in ancient times a herd of these tremendous animals came to the Big Bone licks, and began an universal destruction of the bear, deer, elks, buffaloes, and other animals, which had been created for the use of the Indians; that the Great Man above, looking down

---

From David Waldstreicher, ed., *Notes on the State of Virginia with Related Documents* (New York: Bedford / St. Martin's, 2002), 107–8, 115–16. *Notes on the State of Virginia* is taken from the 1787 English edition of Jefferson's work, which was the first to be published under his name.

and seeing this, was so enraged that he seized his lightning, descended on the earth, seated himself on a neighboring mountain, on a rock, of which his seat and the print of his feet are still to be seen, and hurled his bolts among them till the whole were slaughtered, except the big bull, who, presenting his forehead to the shafts, shook them off as they fell; but missing one at length, it wounded him in the side; whereon, springing round, he bounded over the Ohio, over the Wabash, the Illinois, and finally over the great lakes, where he is living at this day." It is well known that on the Ohio, and in many parts of America further North, tusks, grinders, and skeletons of unparalleled magnitude, are found in great numbers, some lying on the surface of the earth, and some a little below it. A Mr. Stanley, taken prisoner by the Indians near the mouth of the Tanissee, relates that, after being transferred through several tribes, from one to another, he was at length carried over the mountains West of the Missouri to a river which runs westwardly; that these bones abounded there; and that the natives described to him the animal to which they belonged as still existing in the northern parts of their country; from which description he judged it to be an elephant. Bones of the same kind have been lately found some feet below the surface of the earth, in salines opened on the North Holston, a branch of the Tanissee, about the latitude of 36½° North. From the accounts published in Europe, I suppose it to be decided that these are of the same kind with those found in Siberia. . . .

. . . The bones of the mammoth, which have been found in America, are as large as those found in the old world. It may be asked, why I insert the mammoth, as if it still existed? I ask in return why I should omit it, as if it did not exist? Such is the economy of nature, that no instance can be produced of her having permitted any one race of her animals to become extinct; of her having formed any link in her great work so weak as to be broken. To add to this, the traditionary testimony of the Indians, that this animal still exists in the northern and western parts of America, would be adding the light of a taper to that of the meridian sun. Those parts still remain in their aboriginal state, unexplored and undisturbed by us, or by others for us. He may as well exist there now, as he did formerly, where we find his bones. If he be a carnivorous animal, as some anatomists have conjectured, and the Indians affirm, his early retirement may be accounted for from the general destruction of the wild game by the Indians, which commences in the first instant of their connection with us, for the purpose of purchasing matchcoats, hatchets, and fire locks, with their skins.

6

GEORGES CUVIER

Essay on the Theory of the Earth, with Mineralogical Illustrations by Professor Jameson

1822

While Georges Cuvier's great work on the fossil bones of quadrupeds (1812, and another edition in 1821–1824) was too long to encourage full literal translation, his introduction to that work was translated into English. This is a translation of Cuvier by Robert Jameson of the University of Edinburgh, who also provided additional text. It was not pure Cuvier that readers were encountering, but Cuvier reworked by Jameson. Hence Jameson could write in the excerpt that follows, "Cuvier is of opinion that. . . ." Charles Darwin studied with Jameson at the University of Edinburgh in the 1820s and would have been aware of this book. This selection includes a description and an illustration of Megatherium, *historically the most significant of all the fossil quadrupeds. Jameson's translation of Cuvier was important for introducing the subject of species extinction to many English-speaking readers.*

Family. Bruta.

BRADYPUS. SLOTH.

There are but two living species of the sloth tribe, the ai, or bradypus tridactylus; and the unau, or bradypus didactylus. Two fossil species have been described, which are nearly allied not only to these species, but also to the myrmecophaga or ant-eater. The following are the two fossil species:—

1. *Megalonix.*—This remarkable fossil animal appears to have been the size of the ox. Its remains were first discovered in limestone caves in Virginia in the year 1796.

2. *Megatherium.*—This species is the size of the rhinoceros, and its fossil remains have hitherto been found only in South America. The

---

From Georges Cuvier, *Essay on the Theory of the Earth, with Mineralogical Illustrations by Professor Jameson*, 4th ed. (Edinburgh: W. Blackwood, 1822), 370–71, plus plate.

first, and most complete skeleton, was sent from Buenos Ayres by the Marquis Loretto, in the year 1789. It was found in digging an alluvial soil, on the banks of the river Luxan, a league south-east of the village of that name, about three leagues W. S. W. of Buenos Ayres. *Plate 3d* [opposite] gives a faithful representation of this remarkable skeleton, which is now preserved in the Royal Cabinet of Madrid. A second skeleton of the same animal was sent to Madrid from Lima, in the year 1795; and a third was found in Paraguay. Thus it appears, that the remains of this animal exist in the most distant parts of South America. It is very closely allied to the megalonix, and differs from it principally in size, being much larger. Cuvier is of opinion that the two species, the megalonix and megatherium, may be placed together, as members of the same genus, and should be placed between the sloths and ant-eaters, but nearer to the former than to the latter. It is worthy of remark, that the remains of these animals have not been hitherto found in any other quarter of the globe besides America, the only country which affords sloths and ant-eaters.

PLATE III.

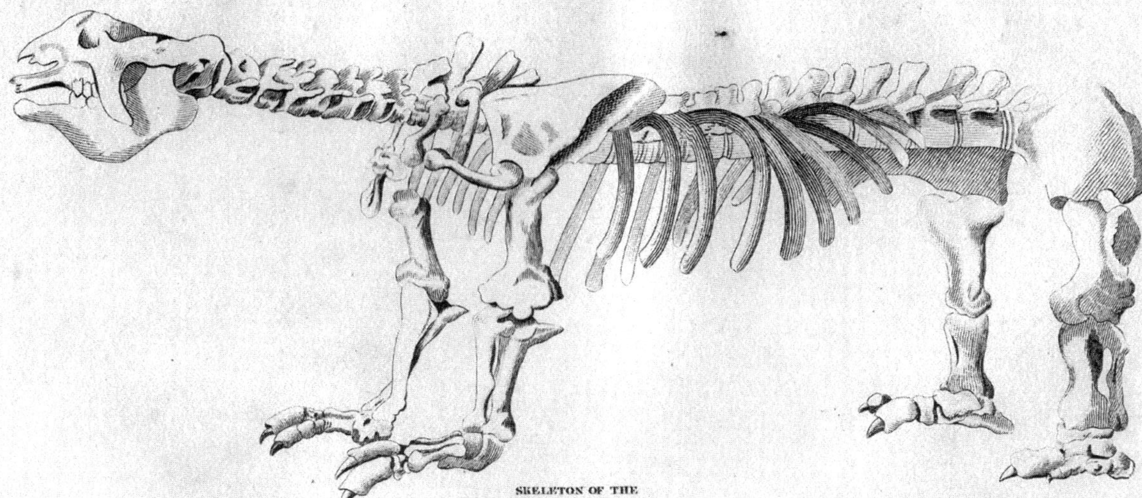

SKELETON OF THE
MEGATHERIUM
DUG OUT of ALLUVIAL STRATA near BUENOS AYRES.

Edinburgh Published by Wm Blackwood 1822

Engd by W. & D. Lizars Edinr

7

THOMAS ROBERT MALTHUS

An Essay on the Principle of Population

1798

Thomas Robert Malthus (1766–1834) was raised in a household where the progressive views of the French and English Enlightenment were the norm. However, as a young man, he reconsidered the liberal views of his parents' generation and came to the conclusion that the prospects for human progress were limited by natural laws governing the growth of population. Malthus's views were enormously influential, both in establishing economics as a social science and for his insight on population, which was exploited by a variety of readers, including Charles Darwin.

The great and unlooked for discoveries that have taken place of late years in natural philosophy; the increasing diffusion of general knowledge from the extension of the art of printing; the ardent and unshackled spirit of inquiry that prevails throughout the lettered, and even unlettered world; the new and extraordinary lights that have been thrown on political subjects, which dazzle, and astonish the understanding; and particularly that tremendous phenomenon in the political horizon the French revolution, which, like a blazing comet, seems destined either to inspire with fresh life and vigour, or to scorch up and destroy the shrinking inhabitants of the earth, have all concurred to lead many able men into the opinion, that we were touching on a period big with the most important changes, changes that would in some measure be decisive of the future fate of mankind. . . .

I have read some of the speculations on the perfectibility of man and of society, with great pleasure. I have been warmed and delighted with the enchanting picture which they hold forth. I ardently wish for such happy improvements. But I see great, and, to my understanding, unconquerable difficulties in the way to them. . . .

From Thomas Robert Malthus, *An Essay on the Principle of Population, as It Affects the Future Improvement of Society, with Remarks on the Speculations of Mr. Godwin, M. Condorcet, and Other Writers* (London: J. Johnson, 1798), 1–2, 7, 11–12, 13–14, 16–17.

I think I may fairly make two postulata.

First, That food is necessary to the existence of man.

Secondly, That the passion between the sexes is necessary, and will remain nearly in its present state.

These two laws ever since we have had any knowledge of mankind, appear to have been fixed laws of our nature; and, as we have not hitherto seen any alteration in them, we have no right to conclude that they will ever cease to be what they now are, without an immediate act of power in that Being who first arranged the system of the universe; and for the advantage of his creatures, still executes, according to fixed laws, all its various operations. . . .

Assuming then, my postulata as granted, I say, that the power of population is indefinitely greater than the power in the earth to produce subsistence for man.

Population, when unchecked, increases in a geometrical ratio. Subsistence increases only in an arithmetical ratio. A slight acquaintance with numbers will shew the immensity of the first power in comparison of the second.

By that law of our nature which makes food necessary to the life of man, the effects of these two unequal powers must be kept equal. . . .

This natural inequality of the two powers of population, and of production in the earth, and that great law of our nature which must constantly keep their effects equal, form the great difficulty that to me appears insurmountable in the way to the perfectibility of society. All other arguments are of slight and subordinate consideration in comparison of this. I see no way by which man can escape from the weight of this law which pervades all animated nature. No fancied equality, no agrarian regulations in their utmost extent, could remove the pressure of it even for a single century. And it appears, therefore, to be decisive against the possible existence of a society, all the members of which, should live in ease, happiness, and comparative leisure; and feel no anxiety about providing the means of subsistence for themselves and families.

Consequently, if the premises are just, the argument is conclusive against the perfectibility of the mass of mankind.

8

WILLIAM PALEY

Natural Theology

1802

William Paley (1743–1805), an Anglican clergyman, wrote this book to draw attention to what he viewed as the evidence of design in nature. The excerpt reproduced here is the famous opening passage where Paley argued from analogy that there was a clear difference between an object such as a stone that showed no evidence of design, and an object such as a watch that did show evidence of design. For the Victorian world, Paley's book represented the classic statement of the argument for intelligent design in nature. Charles Darwin was familiar with William Paley's works, which were an integral part of the curriculum at the University of Cambridge when Darwin was an undergraduate. Both Paley and Darwin attended the university and shared an undergraduate college—Christ's College, where, for a time, Darwin lodged in the same rooms that Paley had occupied as an undergraduate.

In crossing a heath, suppose I pitched my foot against a *stone*, and were asked how the stone came to be there, I might possibly answer, that, for any thing I knew to the contrary, it had lain there for ever: nor would it perhaps be very easy to shew the absurdity of this answer. But suppose I had found a *watch* upon the ground, and it should be enquired how the watch happened to be in that place, I should hardly think of the answer which I had before given, that, for any thing I knew, the watch might have always been there. Yet why should not this answer serve for the watch, as well as for the stone? Why is it not as admissible in the second case, as in the first? For this reason, and for no other, viz. that, when we come to inspect the watch, we perceive (what we could not discover in the stone) that its several parts are framed and put together for a purpose, e.g. that they are so formed and adjusted as to produce motion, and that motion so regulated as to point out the hour

---

From William Paley, *Natural Theology, or Evidences of the Existence and Attributes of the Deity, Collected from the Appearances of Nature* (London: A. Faulder, 1802), 1–4.

of the day; that, if the several parts had been differently shaped from what they are, of a different size from what they are, or placed after any other manner, or in any other order, than that in which they are placed, either no motion at all would have been carried on in the machine, or none which would have answered the use, that is now served by it. To reckon up a few of the plainest of these parts, and of their offices, all tending to one result:—We see a cylindrical box containing a coiled elastic spring, which, by its endeavour to relax itself, turns round the box. We next observe a flexible chain (artificially wrought for the sake of flexure) communicating the action of the spring from the box to the fusee. We then find a series of wheels, the teeth of which catch in, and apply to, each other, conducting the motion from the fusee to the balance, and from the balance to the pointer; and at the same time, by the size and shape of those wheels, so regulating that motion, as to terminate in causing an index, by an equable and measured progression, to pass over a given space in a given time. We take notice that the wheels are made of brass, in order to keep them from rust; the springs of steel, no other metal being so elastic; that over the face of the watch there is placed a glass, a material employed in no other part of the work, but, in the room of which, if there had been any other than a transparent substance, the hour could not be seen without opening the case. This mechanism being observed (it requires indeed an examination of the instrument, and perhaps some previous knowledge of the subject, to perceive and understand it; but being once, as we have said, observed and understood), the inference, we think, is inevitable; that the watch must have had a maker; that there must have existed, at some time and at some place or other, an artificer or artificers who formed it for the purpose which we find it actually to answer; who comprehended its construction, and designed its use.

9

JEAN-BAPTISTE LAMARCK

Zoological Philosophy

1809

This excerpt from Philosophie zoologique *by Jean-Baptiste Lamarck (1744–1829) includes a diagram that offers a concrete clue to his thinking. He imagined that at some point animals sprang directly into existence ("spontaneous generation") with no development from prior forms. He also thought there had to be at least two separate branches of development for animals. In the passage reproduced here, Lamarck also included a theme that he often repeated in his writing concerning the slow passage of time required for evolutionary change. While few of his professional colleagues adopted his views completely, Lamarck's boldness in speculating about the possible course of development in the animal world attracted much interest, and his name remains attached to the idea of evolution.*

The table [opposite] may facilitate the understanding of what I have said. It is there shown that in my opinion the animal scale begins by at least two separate branches, and that as it proceeds it appears to terminate in several twigs in certain places.

This series of animals begins with two branches, where the most imperfect animals are found; the first animals therefore of each of these branches derive existence only through direct or spontaneous generation.

There is one strong reason that prevents us from recognising the successive changes by which known animals have been diversified and been brought to the condition in which we observe them; it is this, that we can never witness these changes. Since we see only the finished work and never see it in course of execution, we are naturally prone to believe that things have always been as we see them rather than that they have gradually developed.

From Jean-Baptiste de Lamarck, *Zoological Philosophy: An Exposition with Regard to the Natural History of Animals*, trans. Hugh Elliot with introductory essays by David L. Hull and Richard W. Burkhardt, Jr. (Chicago: University of Chicago Press, 1984), 178–80.

TABLE

SHOWING THE ORIGIN OF THE VARIOUS ANIMALS.

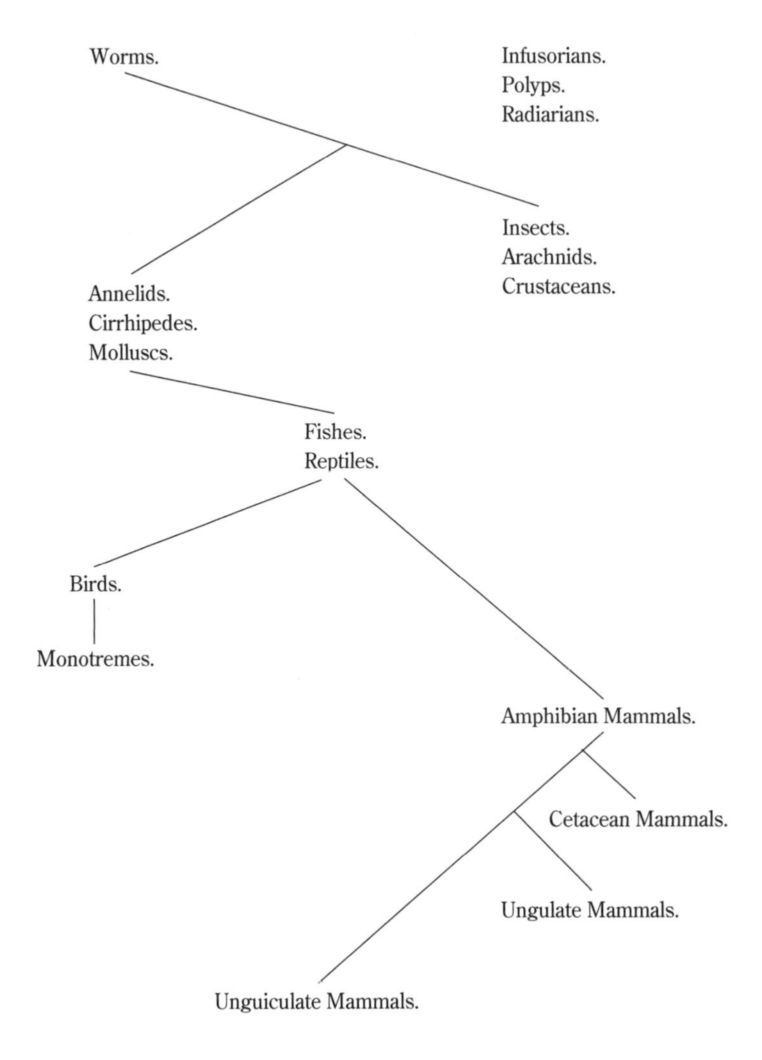

Throughout the changes which nature is incessantly producing in every part without exception, she still remains always the same in her totality and her laws; such changes as do not need a period much longer than the duration of human life are easily recognised by an observer, but he could not perceive any of those whose occurrence consumes a long period of time.

To explain what I mean let me make the following supposition.

If the duration of human life only extended to one second, and if one of our ordinary clocks were wound up and set going, any individual of our species who looked at the hour hand of this clock would detect in it no movement in the course of his life, although the hand is not really stationary. The observations of thirty generations would furnish no clear evidence of a displacement of the hand, for it would only have moved through the distance traversed in half a minute and this would be too small to be clearly perceived; and if still older observations showed that the hand had really changed its position, those who heard this proposition enunciated would not believe it, but would imagine some mistake, since they had always seen the hand at the same point of the dial.

I leave my readers to apply this analogy to the subject in hand.

10

CHARLES LYELL

Principles of Geology

1832

In the second volume of Principles of Geology, *Charles Lyell (1797–1875) took it upon himself to examine the merits of the case for species transformation that had been put forward by Jean-Baptiste Lamarck in his* Zoological Philosophy. *(Lyell had read the work in its original French, which was the preferred language for European science at the time.) Central to Lyell's discussion was answering the question of whether species had a real existence in nature. After examining a number of examples of*

From Charles Lyell, *Principles of Geology, Being an Attempt to Explain the Former Changes of the Earth's Surface, by Reference to Causes Now in Operation*, 3 vols. (London: John Murray, 1830–1833), 2:64–65.

putative change in species, Lyell concluded that the capacity of species to alter was inherent, but only within rather narrow limits. He summarized his anti-Lamarckian conclusions in the following excerpt.

We draw the following inferences, in regard to the reality of *species* in nature.

First, That there is a capacity in all species to accommodate themselves, to a certain extent, to a change of external circumstances, this extent varying greatly according to the species.

2dly. When the change of situation which they can endure is great, it is usually attended by some modifications of the form, colour, size, structure, or other particulars; but the mutations thus superinduced are governed by constant laws, and the capability of so varying forms part of the permanent specific character.

3dly. Some acquired peculiarities of form, structure, and instinct, are transmissible to the offspring; but these consist of such qualities and attributes only as are intimately related to the natural wants and propensities of the species.

4thly. The entire variation from the original type, which any given kind of change can produce, may usually be effected in a brief period of time, after which no farther deviation can be obtained by continuing to alter the circumstances, though ever so gradually,—indefinite divergence, either in the way of improvement or deterioration, being prevented, and the least possible excess beyond the defined limits being fatal to the existence of the individual.

5thly. The intermixture of distinct species is guarded against by the aversion of the individuals composing them to sexual union, or by the sterility of the mule offspring. It does not appear that true hybrid races have ever been perpetuated for several generations, even by the assistance of man; for the cases usually cited relate to the crossing of mules with individuals of pure species, and not to the intermixture of hybrid with hybrid.

6thly. From the above considerations, it appears that species have a real existence in nature, and that each was endowed, at the time of its creation, with the attributes and organization by which it is now distinguished.

11

JOHN HERSCHEL

Letter to Charles Lyell

1836

Charles Lyell sent a copy of the fourth edition (1835) of his Principles of Geology *to John Herschel (1792–1871), the British astronomer who was then making a systematic survey of the stars of the Southern hemisphere from the Cape Observatory at the Cape of Good Hope in Africa. Herschel responded to Lyell's gift with enthusiasm. As this letter indicates, Herschel did not seize on the antitransmutationist conclusion of Lyell's volume but rather on the opening Lyell had provided to consider anew the question of the origin of species. In the 1830s Herschel was probably the most widely admired man of science in England. His designation of the species question as the "mystery of mysteries" confirmed for Lyell that he was pursuing a matter of significance.*

Feldhausen. C. G. H
Feb. 20. *1836*

My dear Sir,

1. I am perfectly ashamed not to have long acknowledged your present of the new Ed. of Geology a work which I now read for the 3d time and every time with increased interest, as it appears to me one of those productions which work a complete revolution in their subject by altering entirely the point of view in which it must thenceforward be contemplated. You have succeeded too in adding dignity to a subject already grand by exposing to view the immense extent & complication of the problems it offers for solution and by unveiling a dim glimpse of a region of speculation connected with it where it seems impossible to venture without experiencing some degree of that mysterious awe which the Sybil appeals to in the bosom of Aeneas on entering the confines of the shades. . . .

From Walter F. [Susan Faye] Cannon, "The Impact of Uniformitarianism: Two Letters from John Herschel to Charles Lyell, 1836–1837," *Proceedings of the American Philosophical Society* 105 (1961): 304–5.

He that on such quest would go must know nor fear nor failing
To coward soul or faithless heart the search were unavailing—

Of course I allude to that mystery of mysteries the replacement of extinct species by others. Many will doubtless think your speculations too bold—but it is as well to face the difficulty at once. For my own part—I cannot but think it an inadequate conception of the Creator, to assume it as granted that his combinations are exhausted upon any one of the theatres of their former exercise—though in this, as in all his other works we are led by all analogy to suppose that he operates through a series of intermediate causes & that in consequence, the origination of fresh species, could it ever come under our cognizance would be found to be a natural in contradistinction to a miraculous process—although we perceive no indications of any process actually in progress which is likely to issue in such a result.

2

Charles Darwin Addresses the Question of Evolution

12

CHARLES DARWIN

Journal of Researches

1839

In 1831, as his university days were coming to an end, Charles Darwin (1809–1882) looked for a way to see the world beyond the shores of Europe. His desire to travel had been whetted by reading Alexander von Humboldt's romantic narrative of his voyage to the New World. When the opportunity arose to join the surveying mission of HMS Beagle, *Darwin leaped at the chance. Like Humboldt, he wanted to delight in the novelty of new and exotic scenery while serving science. During the voyage, Darwin recorded his scientific observations and theories in such subjects as zoology and geology in extensive notes, but he also kept a day-to-day diary of more general experiences and impressions. He regarded his diary not as a record of facts, but of his thoughts.*

Once the Beagle *voyage was over, Captain Robert FitzRoy approached him about publishing their accounts of the voyage in a joint publication, with FitzRoy writing one volume and Darwin another. This was done successfully. At the same time, Darwin's volume was so appealing that the publisher also produced it on its own, under the apt title* Journal of Researches. *What made Darwin's volume so attractive to readers was that he deftly interwove scientific descriptions with his own highly colored*

From Charles Darwin, *Journal of Researches into the Geology and Natural History of the Various Countries Visited by H.M.S. Beagle, 1832–1836* (London: Henry Colburn, 1839), 11, 232–33, 604–5.

and emotive descriptions of scenery. The excerpts that follow include several of these romantic descriptions of scenery. The first excerpt dates from February 1832. It records Darwin's pleasure at wandering by himself in a Brazilian forest. The second excerpt is from the end of the same year. The Beagle *was then in Tierra del Fuego at the tip of South America. As Darwin walked among its hills, he looked south toward the Strait of Magellan, admiring the "mysterious grandeur" of mountain appearing behind mountain. Finally, at the close of the voyage in October 1836, he compared his experiences of Brazil and Tierra del Fuego and added one more memory that stayed with him—that of the Patagonian desert.*

Bahia, or San Salvador. Brazil, Feb. 29th.—The day has past delightfully. Delight itself, however, is a weak term to express the feelings of a naturalist who, for the first time, has been wandering by himself in a Brazilian forest. Among the multitude of striking objects, the general luxuriance of the vegetation bears away the victory. The elegance of the grasses, the novelty of the parasitical plants, the beauty of the flowers, the glossy green of the foliage, all tend to this end. A most paradoxical mixture of sound and silence pervades the shady parts of the wood. The noise from the insects is so loud, that it may be heard even in a vessel anchored several hundred yards from the shore; yet within the recesses of the forest a universal silence appears to reign. To a person fond of natural history, such a day as this, brings with it a deeper pleasure than he ever can hope again to experience. . . .

[Tierra del Fuego.] December 20th.— . . . We followed the same watercourse as on the previous day, till it dwindled away, and then were compelled to crawl blindly among the trees. These, from the effects of the elevation, and of the impetuous winds, were low, thick, and crooked. At length we reached that which from a distance appeared like a carpet of fine green turf, but which, to our vexation, turned out to be a compact mass of little beech-trees about four or five feet high. These were as thick together as box in the border of a flower-garden, and we were obliged to struggle over the flat but treacherous surface. After a little more trouble we gained the peat, and then the bare slate rock.

A ridge connected this hill with another, distant some miles, and more lofty, so that patches of snow were lying on it. As the day was not far advanced, I determined to walk there and collect along the road. It would have been very hard work, had it not been for a well-beaten and straight path made by the guanacoes; for these animals, like sheep, always follow the same line. When we reached the hill we found it the highest in the immediate neighbourhood, and the waters flowed to the sea in opposite directions. We obtained a wide view over the surrounding country: to the northward a swampy moorland extended, but to the southward we had a scene of savage magnificence, well becoming Tierra del Fuego. There was a degree of mysterious grandeur in mountain behind mountain, with the deep intervening valleys, all covered by one thick, dusky mass of forest. The atmosphere, likewise, in this climate (where gale succeeds gale, with rain, hail, and sleet), seems blacker than any where else. In the Strait of Magellan looking due south from Port Famine, the distant channels between the mountains appear from their gloominess to lead beyond the confines of this world. . . .

[Conclusion.] Among the scenes which are deeply impressed on my mind, none exceed in sublimity the primeval forests undefaced by the hand of man; whether those of Brazil, where the powers of Life are predominant, or those of Tierra del Fuego, where Death and Decay prevail. Both are temples filled with the varied productions of the God of Nature:—no one can stand in these solitudes unmoved, and not feel that there is more in man than the mere breath of his body. In calling up images of the past, I find the plains of Patagonia frequently cross before my eyes: yet these plains are pronounced by all most wretched and useless. They are characterized only by negative possessions; without habitations, without water, without trees, without mountains, they support merely a few dwarf plants. Why then, and the case is not peculiar to myself, have these arid wastes taken so firm possession of the memory? Why have not the still more level, the greener and more fertile Pampas, which are serviceable to mankind, produced an equal impression? I can scarcely analyze these feelings: but it must be partly owing to the free scope given to the imagination.

13

[RICHARD OWEN]

Glyptodon clavipes (Gigantic Extinct Armadillo)

1845

The Royal College of Surgeons in England, whose collection was supervised by the comparative anatomist Richard Owen (1804–1892), occasionally published catalogs of its collections. The illustration that follows is of a specimen of Glyptodon clavipes, *originally described by Owen in 1839. The length of the carapace of the specimen is reported as 5 feet 7 inches. Darwin had collected some individual polygonal scutes (plates) from such an animal during the voyage, and he noted that they were larger than but otherwise similar to those of presently living armadillos. Eventually Darwin would interpret present-day armadillos as descendants of some earlier related form such as the* Glyptodon.

Interestingly, Owen labeled the pictured specimen as a "Gigantic Extinct Armadillo," though he did not raise the question of descent in his accompanying descriptive text. Owen was aware of transmutationist theories but did not endorse them, at least not publicly. Owen and Darwin worked closely together during the late 1830s and early 1840s; it was Owen who described the fossil quadrupeds that Darwin brought home with him from the Beagle *voyage. Besides his work in comparative anatomy, Owen is best known for successfully lobbying the British government to establish a museum in London solely devoted to natural history.*

From [Richard Owen], *Descriptive and Illustrated Catalogue of the Fossil Organic Remains of Mammalia and Aves Contained in the Museum of the Royal College of Surgeons of England* (London: Richard and John E. Taylor, 1845), plate I.

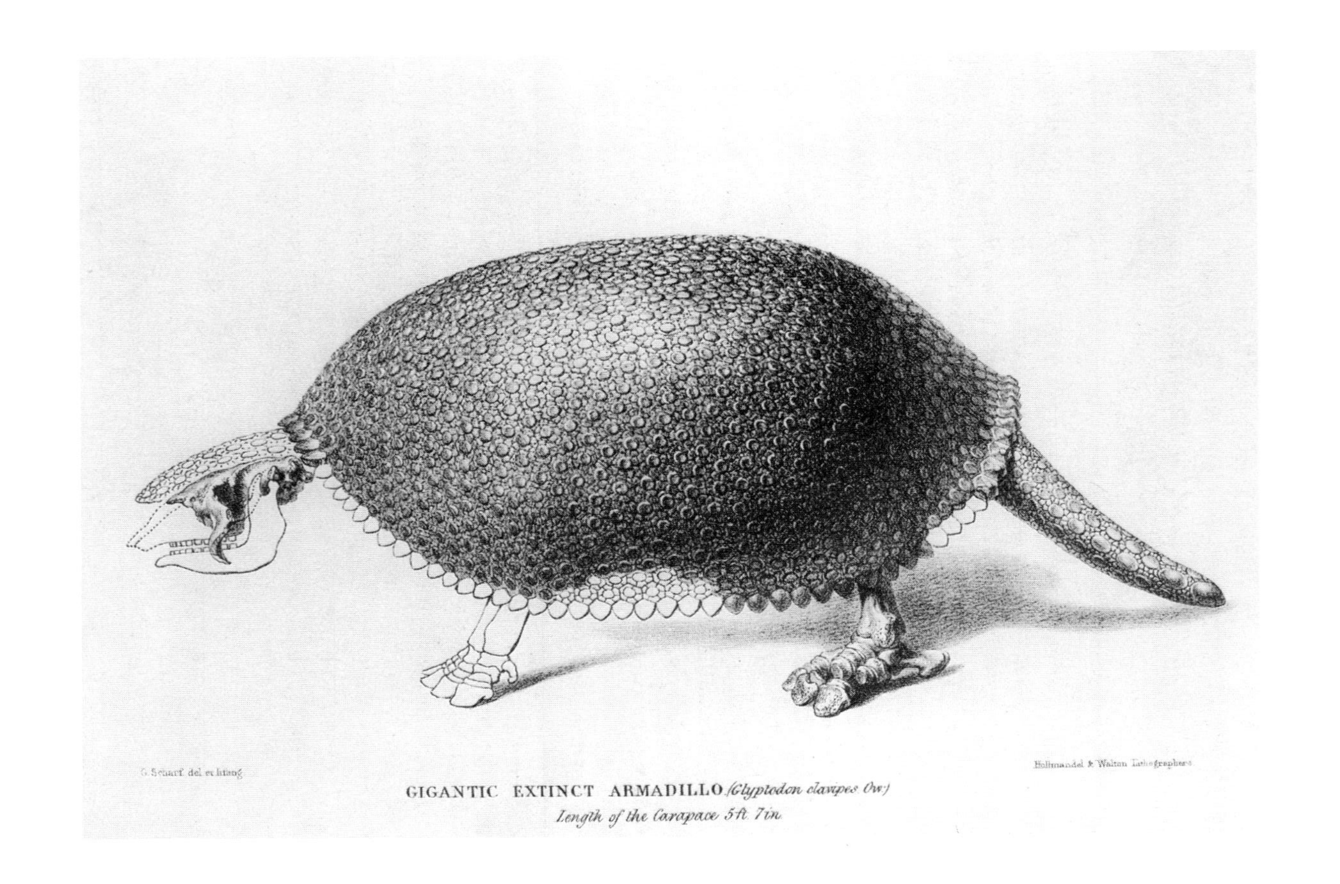

GIGANTIC EXTINCT ARMADILLO *(Glyptodon clavipes Ow)*

Length of the Carapace 5 ft. 7 in

14

CHARLES DARWIN

Ornithological Notes

1836

As HMS Beagle *turned toward home in mid-1836, Charles Darwin made use of the long days at sea to reflect on his observations during the voyage and to organize his collections so that he could disperse them to various specialists immediately upon arriving in England. With the assistance of a copyist, Syms Covington, Darwin drew up lists of the various units of his collections. The excerpt reproduced here is taken from his "Ornithological Notes." It contains one of his earliest general speculations on the species question. The birds he was discussing were all mockingbirds, known by different names in different areas. While visiting the Galápagos Islands in 1835, Darwin had noticed that the mockingbirds varied island by island. After a few months' reflection on their differences, he began to suspect that such facts would "undermine the stability of Species." He was beginning to embrace the idea of species transmutation—or, as he would eventually call it, evolution.*

3306 } Thenca: male: Charles Isd—
3307 } [Thenca: male]: Chatham Isd.—

These birds are closely allied in appearance to the Thenca of Chile (2169) or Callandra of la Plata (1216). In their habits I cannot point out a single difference;—They are lively inquisitive, active *run fast*, frequent houses to pick the meat of the Tortoise, which is hung up,—sing tolerably well; are said to build a simple open nest,—are *very* tame, a character in common with the other birds: I *imagined* however its note or cry was rather different from the Thenca of Chile?—Are very abundant, over the whole Island; are chiefly tempted up into the high & damp parts, by the houses & cleared ground.

I have specimens from four of the larger Islands; the two above enumerated, and (3349: female. Albermarle Isd.) & (3350: male: James Isd).—The

---

From Nora Barlow, ed., "Darwin's Ornithological Notes," *Bulletin of the British Museum (Natural History) Historical Series* 2, no. 7 (1963): 262.

specimens from Chatham & Albermarle Isd appear to be the same; but the other two are different. In each Isld. each kind is *exclusively* found: habits of all are indistinguishable. When I recollect, the fact that the form of the body, shape of scales & general size, the Spaniards can at once pronounce, from which Island any Tortoise may have been brought. When I see these Islands in sight of each other, & [but *del.*] possessed of but a scanty stock of animals, tenanted by these birds, but slightly differing in structure & filling the same place in Nature, I must suspect they are only varieties. The only fact of a similar kind of which I am aware, is the constant asserted difference—between the wolf-like Fox of East & West Falkland Islds.—If there is the slightest foundation for these remarks the zoology of Archipelagoes—will be well worth examining; for such facts [would *inserted*] undermine the stability of Species.

15

CHARLES DARWIN

Notebook B

1837

HMS Beagle *returned to England in October 1836. Darwin quickly unpacked his several thousand zoological specimens and distributed them to experts for identification. Their reports gave him encouragement to believe that species were indeed mutable. Of particular importance was ornithologist John Gould's (1804–1881) judgment that there were three distinct mockingbird species among the Galápagos specimens that Darwin had given him. Darwin interpreted this fact as an indication that the mockingbirds had become altered as they were isolated on the different islands. In consideration of the mockingbird case (and others like it), Darwin became an evolutionist in the spring of 1837. Soon after, he began his first notebook explicitly devoted to the subject. In the following excerpts from the notebook that Darwin labeled "B," he gave free rein to his imagination. His notes were private, rough, and highly speculative. For that reason they provide insight into his thought processes as he tried to understand the nature of species. [In the excerpt, angled double*

From Paul H. Barrett, Peter J. Gautrey, Sandra Herbert, David Kohn, and Sydney Smith, eds., *Charles Darwin's Notebooks, 1836–1844: Geology, Transmutation of Species, Metaphysical Enquiries* (Ithaca, N.Y.: Cornell University Press, 1987), 171, 177, 195.

brackets («/») indicate insertions Darwin made around the time of his initial entry, angled single brackets (‹/›) indicate deletions, and **bold-faced type** *indicates his later annotation.] From 1837 through 1839, Darwin kept a number of such notebooks. Taken together, they provide a record of his path of discovery and reasoning as he devised his theory of evolution through natural selection.*

Why is life short, Why such high object generation.—We *know* world subject to cycle of change, temperature & all circumstances which influence living beings.—

We see ‹living beings›. the young of living beings, become permanently changed or subject to variety, according «to» circumstance,—seeds of plants sown in rich soil, many kinds, are produced, though new individuals produced by buds are constant, hence we see generation here seems a means to vary. or adaptation.—Again we ‹believe› «know» in course of generations even mind & instinct becomes influenced.—child of savage not civilized man.—birds rendered wild ‹through› generations, acquire ideas ditto. V. Zoonomia.—

There may be unknown difficulty with *full grown* individual «with fixed organization» thus being modified,—therefore generation to adapt & alter the race to *changing* world.—

On other hand, generation destroys the effect of accidental injuries, ‹on› which if animals lived for ever would be endless (that is with our present system of body & universe

therefore final cause of life.

. . .

The tree of life should perhaps be called the coral of life, base of branches dead; so that passages cannot be seen.—this again offers contradiction to constant succession of germs in progress.—«no only makes it excessively complicated.»

Is it thus fish can be traced right down to simple organization.—

birds—not.

We may fancy, according to shortness of life of species that in perfection, the bottom of branches deaden.—*so that* in Mammalia «birds» it would only appear like circles;—& insects amongst articulata.—but in lower classes, perhaps a more linear arrangement.—

. . .

The question if creative power acted at Galapagos it so acted that bi[r]ds with plumage «&» tone of voice partly American North & South.— **(& geographical ‹distri› division are arbitrary & not permanent. this might be made very strong. if we believe the Creator creates by any laws. which, I think is shown by the very facts of the Zoological character of these islands** so permanent a breath cannot reside in space before island existed.—Such an influence Must exist in such spots. We know birds do arrive & seeds.—

The same remarks applicable to fossil animals same type, armadillo like covering created.—passage for vertebrae in neck same cause, such beautiful adaptations yet other animals live so well.—This view of propagation gives. ‹ro› hiding place for many unintelligible structures. it might have been of use in progenitor—or it may be of use.—like Mammæ on mens' breasts.—

How does it come wandering birds. such sandpipers. not new at Galapagos.—did the creative force know that «these» species could, arrive—did it only create those kinds not so likely to wander. Did it create two species closely allied to Mus. coronata, but not coronata.—We know that domestic animals vary in countries, without any assignable reason.—

Astronomers might formerly have said that God ordered, each planet to move in its particular destiny.—In same manner God orders each animal created with certain form in certain country, but how much more simple, & sublime power let attraction act according to certain laws such are inevitable consequen let animal be created, then by the fixed laws of generation, such will be their successors.—

16

EMMA DARWIN

Letter to Charles Darwin

c. February 1839

Emma Wedgwood (1808–1896) married her first cousin Charles Darwin in January 1839. She wrote this letter to him shortly after their marriage. The Wedgwood branch of the family was on the whole more traditionally religious than the Darwin branch of the family, though Wedgwoods and Darwins shared a common political orientation, in particular a keen opposition to the institution of slavery. Emma's reference to her new husband as "my own dear Nigger" needs to be read in this light. The person in the letter whom Emma referred to as "E." was Charles's older brother Erasmus Alvey Darwin (1804–1881). The obvious import of the letter has to do with Emma's awareness of her husband's speculations on evolution. While there is no extant letter from Charles to Emma replying to her concerns, Charles wrote at the bottom of Emma's letter, "When I am dead, know that many times, I have kissed & cryed [sic] over this. C.D."

The state of mind that I wish to preserve with respect to you, is to feel that while you are acting conscientiously & sincerely wishing, & trying to learn the truth, you cannot be wrong; but there are some reasons that force themselves upon me & prevent my being always able to give myself this comfort. I dare say you have often thought of them before, but I will write down what has been in my head, knowing that my own dearest will indulge me. Your mind & time are full of the most interesting subjects & thoughts of the most absorbing kind, viz following up yr own discoveries—but which make it very difficult for you to avoid casting out as interruptions other sorts of thoughts which have no relation to what you are pursuing or to be able to give your whole attention to both sides of the question.

There is another reason which would have a great effect on a woman, but I don't know whether it w[d] so much on a man—I mean E. whose understanding you have such a very high opinion of & whom you have

---

From Frederick Burkhardt et al., eds., *The Correspondence of Charles Darwin*, 16+ vols. (Cambridge, U.K.: Cambridge University Press, 1985–), 2:171–73.

so much affection for, having gone before you—is it not likely to have made it easier to you & to have taken off some of that dread & fear which the feeling of doubting first gives & which I do not think an unreasonable or superstitious feeling. It seems to me also that the line of your pursuits may have led you to view chiefly the difficulties on one side, & that you have not had time to consider & study the chain of difficulties on the other, but I believe you do not consider your opinion as formed. May not the habit in scientific pursuits of believing nothing till it is proved, influence your mind too much in other things which cannot be proved in the same way, & which if true are likely to be above our comprehension. I should say also that there is a danger in giving up revelation which does not exist on the other side, that is the fear of ingratitude in casting off what has been done for your benefit as well as for that of all the world & which ought to make you still more careful, perhaps even fearful lest you should not have taken all the pains you could to judge truly. I do not know whether this is arguing as if one side were true & the other false, which I meant to avoid, but I think not. I do not quite agree with you in what you once said—that luckily there were no doubts as to how one ought to act. I think prayer is an instance to the contrary, in one case it is a positive duty & perhaps not in the other. But I dare say you meant in actions which concern others & then I agree with you almost if not quite. I do not wish for any answer to all this—it is a satisfaction to me to write it & when I talk to you about it I cannot say exactly what I wish to say, & I know you will have patience, with your own dear wife. Don't think that it is not my affair & that it does not much signify to me. Every thing that concerns you concerns me & I should be most unhappy if I thought we did not belong to each other forever. I am rather afraid my own dear Nigger will think I have forgotten my promise not to bother him, but I am sure he loves me & I cannot tell him how happy he makes me & how dearly I love him & thank him for all his affection which makes the happiness of my life more & more every day.

17

RODERICK MURCHISON

Presidential Address to the Geological Society of London

February 17, 1843

Fossils were originally valued by geologists as markers of the identity of sedimentary layers of rock on the earth's surface. Such layers, called strata, *were characterized by their fossils. Younger strata lay near the surface of the earth; older strata lay beneath them. However, by the mid-1840s it was increasingly clear that fossils, using the temporal sequence provided by strata, could also be used to reconstruct the history of life on earth. In his presidential address to the Geological Society of London in 1843, geologist Roderick Murchison (1792–1871) used the occasion to present a short history of life on earth.*

Besides ascertaining where the great masses of combustible matter lie, we can now affirm, that during the earliest period of life, conditions prevailed, indicating a prevalence over enormous spaces—if not almost universally—of the same climate, involving a very wide diffusion of similar inhabitants of the ocean. We have learned, that in the earliest of these stages of animal life, no vestige of the vertebrata has yet been found, whilst in the succeeding epochs of the Palæozoic age singular fishes appear, which, in proportion to their antiquity, are more removed from all modern analogies. In each of these early and long-continued periods, the shells preserving on the whole a community of character, differ from each other in each division—and in that later formation, where a very few only of the same types are visible, they are linked on to a new class of beings, the first created of those Saurians, whose existence is prolonged throughout the whole Secondary period; whilst we have this year seen reason to admit that even birds (some of them of gigantic size) may have been the cotemporaries of the first great lizards. With the close of the Palæozoic æra we have also observed a gradual change

---

From Roderick Murchison, "Presidential Address [February 17, 1843]," *Proceedings of the Geological Society of London*, 4 (1843–1845):149–50.

in the plants of the older lands, and that the rank and tropical vegetation of the Carboniferous epoch is succeeded by a peculiar flora. In the next, or Triassic period, we have another flora, whilst new forms of fishes and mollusks indicate an approach to that period when the seas were tenanted by Belemnites and Ammonites, marking so broadly these secondary deposits with which British geologists have long been familiar, and which, commencing with the Lias, terminate with the Chalk. And, lastly, from the dawn of existing races, we ascend through successive deposits gradually becoming more analogous to those of the present day, until at length we reach the bottoms of oceans so recently desiccated, that their shelly remains are undistinguishable from those now associated with Man, the last created in this long chain of animal life in which scarcely a link is wanting!—all bespeaking a perfection and grandeur of design, in contemplating which we are lost in admiration of creative power.

18

[ROBERT CHAMBERS]

Vestiges of the Natural History of Creation

1844

In writing his anonymously published book, Robert Chambers (1802–1871) adopted the schematic history of life that prominent geologists such as Roderick Murchison were providing. To that scheme he added his own interpretation. In Chambers's view, the history of life on earth was a story of progress by steps from fishes on up a line of development that ended with mammals. Chambers was thus in effect reviving the gist of Lamarck's scheme. Unlike Lamarck, however, Chambers sought to place his interpretation of the fossil record in a broader religious and philosophical perspective. For Chambers, the history of life on earth was consistent only with a new understanding of divine Creation, one that emphasized the operation of natural laws rather than the "personal or immediate exertion" on the part of the "Almighty author." The Vestiges*, as it was popularly known, went through many editions both in Britain and in the*

From [Robert Chambers], *Vestiges of the Natural History of Creation* (London: John Churchill, 1844), 148–49, 152–54.

United States and was widely read, not only by men of science but also by the general public. It provoked annoyance among many scientists, who disliked what they regarded as its half-baked schemes. However, the book undoubtedly paved the way for Darwin's Origin of Species, *which was published only fifteen years later.*

In pursuing the progress of the development of both plants and animals upon the globe, we have seen an advance in both cases, along the line leading to the higher forms of organization. Amongst plants, we have first sea-weeds, afterwards land plants; and amongst these the simpler (cellular and cryptogamic) before the more complex. In the department of zoology, we see zoophytes, radiata, mollusca, articulata, existing for ages before there were any higher forms. The first step forward gives fishes, the humblest class of the vertebrata; and, moreover, the earliest fishes partake of the character of the next lowest sub-kingdom, the articulata. Afterwards come land animals, of which the first are reptiles, universally allowed to be the type next in advance from fishes, and to be connected with these by the links of an insensible gradation. From reptiles we advance to birds, and thence to mammalia, which are commenced by marsupialia, acknowledgedly low forms in their class. That there is thus a progress of some kind, the most superficial glance at the geological history is sufficient to convince us. . . .

A candid consideration of all these circumstances can scarcely fail to introduce into our minds a somewhat different idea of organic creation from what has hitherto been generally entertained. That God created animated beings, as well as the terraqueous theatre of their being, is a fact so powerfully evidenced, and so universally received, that I at once take it for granted. But in the particulars of this so highly supported idea, we surely here see cause for some re-consideration. It may now be inquired,—In what way was the creation of animated beings effected? The ordinary notion may, I think, be not unjustly described as this,—that the Almighty author produced the progenitors of all existing species by some sort of personal or immediate exertion. But how does this notion comport with what we have seen of the gradual advance of species, from the humblest to the highest? How can we suppose an immediate exertion of this creative power at one time to produce zoophytes, another time to add a few marine mollusks, another to bring in one or two conchifers,[1] again to produce crustaceous fishes, again perfect fishes, and

[1] *conchifers*: A bivalve mollusc of the class *Conchifera*.

so on to the end? This would surely be to take a very mean view of the Creative Power—to, in short, anthropomorphize it, or reduce it to some such character as that borne by the ordinary proceedings of mankind. And yet this would be unavoidable; for that the organic creation was thus progressive through a long space of time, rests on evidence which nothing can overturn or gainsay. Some other idea must then be come to with regard to *the mode* in which the Divine Author proceeded in the organic creation. Let us seek in the history of the earth's formation for a new suggestion on this point. We have seen powerful evidence, that the construction of this globe and its associates, and inferentially that of all the other globes of space, was the result, not of any immediate or personal exertion on the part of the Deity, but of natural laws which are expressions of his will. What is to hinder our supposing that the organic creation is also a result of natural laws, which are in like manner an expression of his will? More than this, the fact of the cosmical arrangements being an effect of natural law, is a powerful argument for the organic arrangements being so likewise, for how can we suppose that the august Being who brought all these countless worlds into form by the simple establishment of a natural principle flowing from his mind, was to interfere personally and specially on every occasion when a new shell-fish or reptile was to be ushered into existence on *one* of these worlds? Surely this idea is too ridiculous to be for a moment entertained.

19

ALFRED RUSSEL WALLACE

On the Law Which Has Regulated the Introduction of New Species

1855

Alfred Russel Wallace (1823–1913) wrote this article in February 1855 from Sarawak, Borneo, where he was doing research. While it was not overtly evolutionary, its recognition of pattern and regularity in the

From Alfred Russel Wallace, "On the Law Which Has Regulated the Introduction of New Species," *Annals and Magazine of Natural History* 16 (1815): 185–86.

introduction of species marked an advance in the debate over species. The article, which was widely read, brought Wallace to the attention of Charles Lyell and Charles Darwin.

The following propositions in Organic Geography and Geology give the main facts on which the hypothesis is founded.

Geography

1. Large groups, such as classes and orders, are generally spread over the whole earth, while smaller ones, such as families and genera, are frequently confined to one portion, often to a very limited district.
2. In widely distributed families the genera are often limited in range; in widely distributed genera, well-marked groups of species are peculiar to each geographical district.
3. When a group is confined to one district, and is rich in species, it is almost invariably the case that the most closely allied species are found in the same locality or in closely adjoining localities, and that therefore the natural sequence of the species by affinity is also geographical.
4. In countries of a similar climate, but separated by a wide sea or lofty mountains, the families, genera and species of the . . . one are often represented by closely allied families, genera and species peculiar to the other.

Geology

5. The distribution of the organic world in time is very similar to its present distribution in space.
6. Most of the larger and some small groups extend through several geological periods.
7. In each period, however, there are peculiar groups, found nowhere else, and extending through one or several formations.
8. Species of one genus, or genera of one family occurring in the same geological time are more closely allied than those separated in time.
9. As generally in geography no species or genus occurs in two very distant localities without being also found in intermediate

places, so in geology the life of a species or genus has not been interrupted. In other words, no group or species has come into existence twice.

10. The following law may be deduced from these facts:—*Every species has come into existence coincident both in space and time with a pre-existing closely allied species.*

20

CHARLES DARWIN

Letter to Asa Gray

1857

On September 5, 1857, Darwin wrote to American botanist Asa Gray (1810–1888) asking for his opinion regarding a sketch of the theory of evolution that he was then in the process of writing up. Darwin sent the passage that appears below as an attachment to the letter. The material in the attachment turned out to be important not only in persuading Gray of the merits of the argument but also as evidence establishing Darwin's priority when, the next year, Wallace's letter containing similar ideas arrived on Darwin's desk. The material below was eventually published as one of the documents in Charles Darwin and Alfred Russel Wallace, "On the Tendency of Species to Form Varieties; and on the Perpetuation of Varieties and Species by Natural Means of Selection" (Read July 1, 1858), Journal of the Linnean Society of London (Zoology)*, 3 (1859): 45–62. Neither Darwin nor Wallace was present on July 1, 1858, at the Linnean Society.*

I. It is wonderful what the principle of Selection by Man, that is the picking out of individuals with any desired quality, and breeding from them, and again picking out, can do. Even Breeders have been astonished at their own results. They can act on differences inappreciable to

From Frederick Burkhardt et al., eds., *The Correspondence of Charles Darwin*, 16+ vols. (Cambridge, U.K.: Cambridge University Press, 1985–), 6:447–50; also see 7:507–11 for a transcription of Darwin's original draft of the attachment.

an uneducated eye. Selection has been *methodically* followed in *Europe* for only the last half century. But it has occasionally, and even in some degree methodically, been followed in the most ancient times. There must have been, also, a kind of unconscious selection from the most ancient times,—namely in the preservation of the individual animals (without any thought of their offspring) most useful to each race of man in his particular circumstances. The "rogueing" as nurserymen call the destroying of varieties, which depart from their type, is a kind of selection. I am convinced that intentional and occasional selection has been the main agent in making our domestic races. But, however, this may be, its great power of modification has been indisputably shown in late times. Selection acts only by the accumulation of very slight or greater variations, caused by external conditions, or by the mere fact that in generation the child is not absolutely similar to its parent. Man by this power of accumulating variations adapts living beings to his wants,—he *may be said* to make the wool of one sheep good for carpets and another for cloth &c.—

II. Now suppose there was a being, who did not judge by mere external appearance, but could study the whole internal organization—who never was capricious,—who should go on selecting for one end during millions of generations, who will say what he might not effect! In nature we have some *slight* variations, occasionally in all parts: and I think it can be shown that a change in the conditions of existence is the main cause of the child not exactly resembling its parents; and in nature geology shows us what changes have taken place, and are taking place. We have almost unlimited time: no one but a practical geologist can fully appreciate this: think of the Glacial period, during the whole of which the same species of shells at least have existed; there must have been during this period millions on millions of generations.

III. I think it can be shown that there is such an unerring power at work, or *Natural Selection* (the title of my Book), which selects exclusively for the good of each organic being. The elder De Candolle, W. Herbert, and Lyell have written strongly on the struggle for life; but even they have not written strongly enough. Reflect that every being (even the Elephant) breeds at such a rate, that in a few years, or at most a few centuries or thousands of years, the surface of the earth would not hold the progeny of any one species. I have found it hard constantly to bear in mind that the increase of every single species is checked during some part of its life, or during some shortly recurrent generation. Only a few of those annually born can live to propagate their kind. What a trifling difference must often determine which shall survive and which perish—

IV. Now take the case of a country undergoing some change; this will tend to cause some of its inhabitants to vary slightly; not but what I believe most beings vary at all times enough for selection to act on. Some of its inhabitants will be exterminated, and the remainder will be exposed to the mutual action of a different set of inhabitants, which I believe to be more important to the life of each being than mere climate. Considering the infinitely various ways, beings have to obtain food by struggling with other beings, to escape danger at various times of life, to have their eggs or seeds disseminated &c. &c, I cannot doubt that during millions of generations individuals of a species will be born with some slight variation profitable to some part of its economy; such will have a better chance of surviving, propagating, this variation, which again will be slowly increased by the accumulative action of Natural selection; and the variety thus formed will either coexist with, or more commonly will exterminate its parent form. An organic being like the woodpecker or misletoe may thus come to be adapted to a score of contingencies: natural selection, accumulating those slight variations in all parts of its structure which are in any way useful to it, during any part of its life.

V. Multiform difficulties will occur to everyone on this theory. Most can I think be satisfactorily answered.—"Natura non facit saltum" ["Nature makes no leap"] answers some of the most obvious.—The slowness of the change, and only a very few undergoing change at any one time answers others. The extreme imperfections of our geological records answers others.—

VI. One other principle, which may be called the principle of divergence plays, I believe, an important part in the origin of species. The same spot will support more life if occupied by very diverse forms: we see this in the many generic forms in a square yard of turf (I have counted 20 species belonging to 18 genera),—or in the plants and insects, on any little uniform islet, belonging almost to as many genera and families as to species.—We can understand this with the higher, animals whose habits we best understand. We know that it has been experimentally shown that a plot of land will yield a greater weight, if cropped with several species of grasses than with 2 or 3 species. Now every single organic being, by propagating so rapidly, may be said to be striving its utmost to increase in numbers. So it will be with the offspring of any species after it has broken into varieties or sub-species or true species. And it follows, I think, from the foregoing facts that the varying offspring of each species will try (only few will succeed) to seize on as many and as diverse places in the economy of nature, as possible. Each new variety or species, when formed will generally take the places of and so exterminate its less well-fitted parent. This, I believe, to be the origin of the classification or arrangement of all organic beings at all times. These always **seem** to branch and sub-branch like a tree from a common trunk; the flourishing twigs destroying the less vigorous,—the dead and lost branches rudely representing extinct genera and families.

This sketch is *most* imperfect; but in so short a space I cannot make it better. Your imagination must fill up many wide blanks.—Without some reflexion it will appear all rubbish; perhaps it will appear so after reflexion.—| C. D.

21

ALFRED RUSSEL WALLACE

Recollections

1858

Alfred Russel Wallace published an autobiography in 1905. In it he recalled an important moment in his life that had occurred decades earlier. The moment occurred in 1858 when Wallace was on the island of Ternate, now part of Indonesia. He had been engaged in collecting species and studying their patterns of geographic distribution. Wallace was already convinced that evolution had occurred, but he did not have any idea how it had occurred. His work on Ternate was difficult largely because he suffered from malaria. The intermittent fevers caused by the disease often forced him to rest during the day. On one such day, while pondering the subject of species, he recalled reading Thomas Robert Malthus's Essay on Population *twelve years earlier. Malthus's ideas on population led Wallace "to ask the question, Why do some die and some live?" Wallace realized "on the whole the best fitted live." He then wrote to Charles Darwin to describe his new idea. He had been impressed with Darwin's writings on species, but did not realize that Darwin had already come to a similar idea, also drawn from a reading of Malthus. Wallace is credited today as the coauthor, with Charles Darwin, of the theory of evolution through natural selection, but he is not as well known as Darwin.*

From Alfred Russel Wallace, *My Life: A Record of Events and Opinions*, 2 vols. (New York: Dodd, Mead, 1905), 1:361–63.

At the time in question I was suffering from a sharp attack of intermittent fever, and every day during the cold and succeeding hot fits had to lie down for several hours, during which time I had nothing to do but to think over any subjects then particularly interesting me. One day something brought to my recollection Malthus's "Principles of Population," which I had read about twelve years before. I thought of his clear exposition of "the positive checks to increase"—disease, accidents, war, and famine—which keep down the population of savage races to so much lower an average than that of more civilized peoples. It then occurred to me that these causes or their equivalents are continually acting in the case of animals also; and as animals usually breed much more rapidly than does mankind, the destruction every year from these causes must be enormous in order to keep down the numbers of each species, since they evidently do not increase regularly from year to year, as otherwise the world would long ago have been densely crowded with those that breed most quickly. Vaguely thinking over the enormous and constant destruction which this implied, it occurred to me to ask the question, Why do some die and some live? And the answer was clearly, that on the whole the best fitted live. From the effects of disease the most healthy escaped; from enemies, the strongest, the swiftest, or the most cunning; from famine, the best hunters or those with the best digestion; and so on. Then it suddenly flashed upon me that this self-acting process would necessarily *improve the race,* because in every generation the inferior would inevitably be killed off and the superior would remain—that is, *the fittest would survive.* Then at once I seemed to see the whole effect of this, that when changes of land and sea, or of climate, or of food-supply, or of enemies occurred—and we know that such changes have always been taking place—and considering the amount of individual variation that my experience as a collector had shown me to exist, then it followed that all the changes necessary for the adaptation of the species to the changing conditions would be brought about; and as great changes in the environment are always slow, there would be ample time for the change to be effected by the survival of the best fitted in every generation. In this way every part of an animal's organization could be modified exactly as required, and in the very process of this modification the unmodified would die out, and thus the *definite* characters and the clear *isolation* of each new species would be explained. The more I thought over it the more I became convinced that I had at length found the long-sought-for law of nature that solved the problem of the origin of species. For the next hour I thought over the deficiencies in the theories of Lamarck and of the author of the "Vestiges," and I saw that my new theory supplemented these views and obviated every important difficulty. I waited anxiously for the termination of my fit so that I might at once make notes for a paper on the subject. The same evening I did this pretty fully, and on the two succeeding evenings wrote it out carefully in order to send it to Darwin by the next post, which would leave in a day or two.

I wrote a letter to him in which I said that I hoped the idea would be as new to him as it was to me, and that it would supply the missing factor to explain the origin of species.

22

CHARLES DARWIN

Recollections

1831–1858

Between the ages of 67 and 73, Charles Darwin wrote what he called "Recollections of the Development of My Mind and Character." These recollections were not published during his lifetime, though an expurgated version was published by his son Francis Darwin in 1887. A full version of the manuscript, with original omissions restored, was published by his granddaughter Nora Barlow in 1958. In this selection, Darwin reflects on the history of his work on the species question from the time of the voyage of the Beagle *(1831–1836) through his receipt in 1858 of Alfred Russel Wallace's letter describing a similar theory of evolution through natural selection. After receiving Wallace's letter, Darwin turned it over to Charles Lyell and Joseph Hooker, who recommended joint publication of the theory. Thus it was that the first presentation of the theory of evolution through natural selection occurred at a meeting of the Linnean Society of London on July 1, 1858. Neither Darwin nor Wallace was present at the meeting—Darwin because of chronic illness, Wallace because he was still overseas. Portions of their writings on the subject of evolution were read aloud in their absence. It is interesting to compare Wallace's*

From Francis Darwin, ed., *The Life and Letters of Charles Darwin, including an Autobiographical Chapter*, vol. 1 (London: John Murray, 1887), 82–85.

account of his arrival at the idea of natural selection (Document 21) with Darwin's.

During the voyage of the *Beagle* I had been deeply impressed by discovering in the Pampean formation great fossil animals covered with armour like that on the existing armadillos; secondly, by the manner in which closely allied animals replace one another in proceeding southwards over the Continent; and thirdly, by the South American character of most of the productions of the Galapagos archipelago, and more especially by the manner in which they differ slightly on each island of the group; none of the islands appearing to be very ancient in a geological sense.

It was evident that such facts as these, as well as many others, could only be explained on the supposition that species gradually become modified; and the subject haunted me. But it was equally evident that neither the action of the surrounding conditions, nor the will of the organisms (especially in the case of plants) could account for the innumerable cases in which organisms of every kind are beautifully adapted to their habits of life—for instance, a woodpecker or a tree-frog to climb trees, or a seed for dispersal by hooks or plumes. I had always been much struck by such adaptations, and until these could be explained it seemed to me almost useless to endeavour to prove by indirect evidence that species have been modified.

After my return to England it appeared to me that by following the example of Lyell in Geology, and by collecting all facts which bore in any way on the variation of animals and plants under domestication and nature, some light might perhaps be thrown on the whole subject. My first note-book was opened in July 1837. I worked on true Baconian principles, and without any theory collected facts on a wholesale scale, more especially with respect to domesticated productions, by printed enquiries, by conversation with skilful breeders and gardeners, and by extensive reading. When I see the list of books of all kinds which I read and abstracted, including whole series of Journals and Transactions, I am surprised at my industry. I soon perceived that selection was the keystone of man's success in making useful races of animals and plants. But how selection could be applied to organisms living in a state of nature remained for some time a mystery to me.

In October 1838, that is, fifteen months after I had begun my systematic enquiry, I happened to read for amusement "Malthus on Population," and being well prepared to appreciate the struggle for existence which everywhere goes on from long-continued observation of the habits of animals and plants, it at once struck me that under these circumstances favourable variations would tend to be preserved, and unfavourable ones to be destroyed. The result of this would be the formation of new species. Here then I had at last got a theory by which to work; but I was so anxious to avoid prejudice, that I determined not for some time to write even the briefest sketch of it. In June 1842 I first allowed myself the satisfaction of writing a very brief abstract of my theory in pencil in 35 pages; and this was enlarged during the summer of 1844 into one of 230 pages, which I had fairly copied out and still possess.

But at that time I overlooked one problem of great importance; and it is astonishing to me, except on the principle of Columbus and his egg,[1] how I could have overlooked it and its solution. This problem is the tendency in organic beings descended from the same stock to diverge in character as they become modified. That they have diverged greatly is obvious from the manner in which species of all kinds can be classed under genera, genera under families, families under sub-orders and so forth; and I can remember the very spot in the road, whilst in my carriage, when to my joy the solution occurred to me; and this was long after I had come to Down. The solution, as I believe, is that the modified offspring of all dominant and increasing forms tend to become adapted to many and highly diversified places in the economy of nature.

Early in 1856 Lyell advised me to write out my views pretty fully, and I began at once to do so on a scale three or four times as extensive as that which was afterwards followed in my "Origin of Species"; yet it was only an abstract of the materials which I had collected, and I got through about half the work on this scale. But my plans were overthrown, for early in the summer of 1858 Mr. Wallace, who was then in the Malay archipelago, sent me an essay "On the Tendency of Varieties to depart indefinitely from the Original Type"; and this essay contained exactly the same theory as mine. Mr. Wallace expressed the wish that if I thought well of his essay, I should send it to Lyell for perusal.

[1] A story whose moral is that once one knows how to do something it is easy.

23

WHITWELL ELWIN

Letter to John Murray

May 3, 1859

After he received Alfred Russel Wallace's 1858 letter, Darwin abandoned his long manuscript on species, which he had been working on since 1856, in favor of a much shorter work that was eventually published as On the Origin of Species. *In shortening his manuscript, Darwin eliminated many of the examples that had illustrated various points in his argument. He also placed the theoretical portion of his argument prominently in the opening chapters. Emphasizing theory over fact was a new approach in writing on natural history. Darwin knew that his approach would mark a departure from the customary format of books on natural history, but he was determined to express his views succinctly and accurately.*

To do this he had to have the approval of his publisher, John Murray (1808–1892). Before going ahead with the shorter manuscript, Murray sent the manuscript out for review to Whitwell Elwin (1816–1900), editor of the Quarterly Review, *one of the most important journals of opinion in London. As is clear from Elwin's letter of May 3, he was fairly uncomfortable with the heavily theoretical nature of Darwin's book. He suggested, as an alternative, that Darwin place his studies of variation in pigeons in the central position of the book. As Elwin noted, "Every body is interested in pigeons." Elwin was hoping that the new manuscript would prove to be as "charming" a book as Darwin's* Journal of Researches *from the* Beagle *voyage. Elwin's views no doubt reflected what was the majority opinion among those who commonly read books in natural history. In the end, Darwin got his way and published the book he wanted to publish: The* Origin *is a highly theoretical work, albeit with substantial amounts of factual evidence included. However, in subsequent works, Darwin published empirical research that had led him to his views and that, for reasons of economy of presentation, he had omitted from the* Origin.

From Frederick Burkhardt et al., eds., *The Correspondence of Charles Darwin*, 16+ vols. (Cambridge, U.K.: Cambridge University Press, 1985–), 7:288–91.

My dear Murray,

I have been intending for some days to write to you upon the subject of Mr. Darwin's work on the Origin of Species. After you had the kindness to allow me to read the Ms. I made a point of seeing Sir C. Lyell, who I understood had, in some degree, advised the publication. I had myself formed a strong opinion the other way, & I stated to him fully my conviction, & the grounds of it. When we had thoroughly talked the matter over Sir Charles considered that I ought through you to convey my impressions to Mr. Darwin himself. I should have thought this presumptuous & impertinent in me if I had not received from Sir Charles the assurance that Mr. Darwin would not consider it either the one or the other. Nevertheless I speak with diffidence, & am sorry that Sir Charles, who was just starting for the continent, could not, before his return, find leisure to correspond with Mr. Darwin on the question.

I must say at the outset that it is the very high opinion I have of Mr. Darwin, founded on his Journal of a Naturalist, & the conviction, amounting to certainty, of the value of any researches of his, which made me eager to get both him & his friends to re-consider the propriety of sending forth his treatise in its present form. It seemed to me that to put forth the theory without the evidence would do grievous injustice to his views, & to his twenty years of observation & experiment. At every page I was tantalised by the absence of the proofs. All kinds of objections, & possibilities rose up in the mind, & it was fretting to think that the author had a whole array of facts, & inferences from the facts, absolutely *essential* to the decision of the question which were not before the reader. It is to ask the jury for a verdict without putting the witnesses into the box. One part of the public I suspect, under these circumstances, will reject the theory from recalling some obvious facts apparently at variance with it, & to which Mr. Darwin may nevertheless have a complete answer, while another part of the public will feel how unsatisfactory it is to go into the theory when only a fragment of the subject is before them, & will postpone the consideration of it till they can study it with more advantage. The more original the view, the more elaborate the researches on which it rests, the more extensive the series of facts in Natural History which bear upon it, the more it is prejudiced by a partial survey of the field which keeps out of sight the larger part of the materials.

A second objection to the publication of the treatise in its present form, though of less weight than the first, is yet of some moment. The Journal of Mr. Darwin is, as you have often heard me say, one of the most charming books in the language. No person could detail observations in natural history in a more attractive manner. The dissertation on species is, on the contrary, in a much harder & drier style. I impute this to the absence of the details. It is these which give relief & interest to the scientific outline—so that the very omission which takes from the philosophical value of the work destroys in a great degree its popular value also. Whatever class of the public he wishes to win he weakens the effect by an imperfect, & comparatively meagre exposition of his theory.

I am aware that many facts are given in the work as it stands, but they are too often wanting to do more than qualify my criticisms. I state my views broadly & roughly. Mr. Darwin will understand my meaning as well as if I had spoken with nice precision.

Upon the supposition that my description of the work is correct Sir C. Lyell agrees in my conclusions & bid me say this when I wrote you a letter for Mr. Darwin to read. Sir Charles tells me that he feared that in his anxiety to make his work perfect Mr. Darwin would postpone indefinitely the putting his materials into shape, & that thus the world might at last be deprived of his labours. He also told me that another gentleman had put forward a similar theory, & that it was necessary that Mr. D. should promulgate his conclusions before he was anticipated. Influenced by these considerations Sir Charles urged the publication of Mr. D's observations upon pigeons, which he informs me are curious, ingenious, & valuable in the highest degree, accompanied with a brief statement of his general principles. He might then remark that of these principles the phenomena respecting the pigeons were one illustration, & that a larger work would shortly appear in which the same conclusions would be demonstrated by examples drawn from the wide world of nature.

This appears to me to be an admirable suggestion. Even if the larger work were ready it would be the best mode of preparing the way for it. Every body is interested in pigeons. The book would be reviewed in every journal in the kingdom, & would soon be on every table. The public at large can better understand a question when it is narrowed to a single case of this kind than when the whole varied kingdom of nature is brought under discussion at the outset. Interest in the larger work would be roused, & good-will would be conciliated to the subsequent development of the theory in all its bearings. It would be approached with impartiality,—not to say favour, & would appeal to the large public which had been interested by the previous book upon pigeons,—which book would yet be complete in itself, & open to none of the objections that I have urged against the present outline. Indeed I should say of the latter that for an outline it is too much, & for a thorough discussion of the question it is not near enough.

I write this letter with the intention that you should forward it to Mr. Darwin. He must be good enough to excuse the crude manner in which I state my impressions. I am obliged to write as fast as my pen can move, or I should not be able to write at all. My sole object & desire is to secure his theory coming before the world in the way which will do justice to the extraordinary merit of his investigations, & procure him that fame which belongs to him. I am but a smatterer in these subjects. What I say has no sort of authority except so far as it may chance to recommend itself to Mr. Darwin's own reason. The book on pigeons would be at any rate a delightful commencement & I am certain its reception would be the best stimulus to the prosecution of his subsequent work. I should hope if he inclines to this view that the preparatory volume could soon be got ready for the press.

Believe me | Most Sincerely yrs | Whitwell Elwin.

24

CHARLES DARWIN

On the Origin of Species by Means of Natural Selection, or the Preservation of Favoured Races in the Struggle for Life

1859

The four excerpts that follow are taken from the Origin. *The first excerpt contains the title page and facing page and the table of contents—called the* front matter *in the book trade. The front matter offers a great deal of information about the content and overall orientation of the book, particularly in nineteenth-century books where the custom was to embellish the title page (or the one facing it) with relevant epigraphs and to summarize the contents of each chapter in detail in the table of contents. Darwin's* Origin *is a good example of a Victorian book. The epigraphs by William Whewell and Francis Bacon emphasize the rule of natural law in the material world. The full title of the book is long and descriptive (the word* races *is equivalent to the present-day term* varieties*). Darwin is*

From Charles Darwin, *On the Origin of Species by Means of Natural Selection, or the Preservation of Favoured Races in the Struggle for Life* (London: John Murray, 1859), title page and facing page, v–ix, 20–21, 23, 90–91, 484–90.

identified by his membership in various scientific bodies and as the author of the Journal of Researches *from the* Beagle *voyage. There is a strong contrast here between the* Vestiges of the Natural History of Creation *(Document 18) and the* Origin *on this point.*

In the table of contents, Darwin's fourteen chapters are summarized. The essential part of the argument of the book is contained in its first four chapters. The excerpt taken from Chapter I illustrates Darwin's interest in domestic pigeons as an instance of artificial selection. The excerpt from Chapter IV pertains to natural selection, and the excerpt from Chapter XIV discusses the future of evolutionary ideas. Although usually modest in his manner of self-expression, Darwin predicts his book would mark a "considerable revolution in natural history."

> *"But with regard to the material world, we can at least go so far as this—we can perceive that events are brought about not by insulated interpositions of Divine power, exerted in each particular case, but by the establishment of general laws."* —W. WHEWELL: *Bridgewater Treatise.*

> *"To conclude, therefore, let no man out of a weak conceit of sobriety, or an ill-applied moderation, think or maintain, that a man can search too far or be too well studied in the book of God's word, or in the book of God's works; divinity or philosophy; but rather let men endeavour an endless progress or proficience in both."* —BACON: *Advancement of Learning.*

ON

THE ORIGIN OF SPECIES

BY MEANS OF NATURAL SELECTION,

OR THE

PRESERVATION OF FAVOURED RACES IN THE STRUGGLE
FOR LIFE.

By CHARLES DARWIN, M.A.,

FELLOW OF THE ROYAL, GEOLOGICAL, LINNÆAN, ETC., SOCIETIES;
AUTHOR OF 'JOURNAL OF RESEARCHES DURING H. M. S. BEAGLE'S VOYAGE
ROUND THE WORLD.'

LONDON:

JOHN MURRAY, ALBEMARLE STREET.

1859.

CONTENTS.

[From Chapter I]

On the Breeds of the Domestic Pigeon.—Believing that it is always best to study some special group, I have, after deliberation, taken up domestic pigeons. I have kept every breed which I could purchase or obtain, and have been most kindly favoured with skins from several quarters of the world, more especially by the Hon. W. Elliot from India, and by the Hon. C. Murray from Persia. Many treatises in different languages have been published on pigeons, and some of them are very important, as being of considerable antiquity. I have associated with several eminent fanciers, and have been permitted to join two of the London Pigeon Clubs. The diversity of the breeds is something astonishing. . . .

Great as the differences are between the breeds of pigeons, I am fully convinced that the common opinion of naturalists is correct, namely, that all have descended from the rock-pigeon (Columba livia), including under this term several geographical races or sub-species, which differ from each other in the most trifling respects.

[From Chapter IV]

Illustrations of the action of Natural Selection.—In order to make it clear how, as I believe, natural selection acts, I must beg permission to give one or two imaginary illustrations. Let us take the case of a wolf, which preys on various animals, securing some by craft, some by strength, and some by fleetness; and let us suppose that the fleetest prey, a deer for instance, had from any change in the country increased in numbers, or that other prey had decreased in numbers, during that season of the year when the wolf is hardest pressed for food. I can under such circumstances see no reason to doubt that the swiftest and slimmest wolves would have the best chance of surviving, and so be preserved or selected,—provided always that they retained strength to master their prey at this or at some other period of the year, when they might be compelled to prey on other animals. I can see no more reason to doubt this, than that man can improve the fleetness of his greyhounds by careful and methodical selection, or by that unconscious selection which results from each man trying to keep the best dogs without any thought of modifying the breed.

[From Chapter XIV]

When the views entertained in this volume on the origin of species, or when analogous views are generally admitted, we can dimly foresee that there will be a considerable revolution in natural history. Systematists will be able to pursue their labours as at present; but they will not be incessantly haunted by the shadowy doubt whether this or that form be in essence a species. This I feel sure, and I speak after experience, will be no slight relief. The endless disputes whether or not some fifty species of British brambles are true species will cease. Systematists will have only to decide (not that this will be easy) whether any form be sufficiently constant and distinct from other forms, to be capable of definition; and if definable, whether the differences be sufficiently important to deserve a specific name. This latter point will become a far more essential consideration than it is at present; for differences, however slight, between any two forms, if not blended by intermediate gradations, are looked at by most naturalists as sufficient to raise both forms to the rank of species. Hereafter we shall be compelled to acknowledge that the only distinction between species and well-marked varieties is, that the latter are known, or believed, to be connected at the present day by intermediate gradations, whereas species were formerly thus connected. Hence, without quite rejecting the consideration of the present existence of intermediate gradations between any two forms, we shall be led to weigh more carefully and to value higher the actual amount of difference between them. It is quite possible that forms now generally acknowledged to be merely varieties may hereafter be thought worthy of specific names, as with the primrose and cowslip; and in this case scientific and common language will come into accordance. In short, we shall have to treat species in the same manner as those naturalists treat genera, who admit that genera are merely artificial combinations made for convenience. This may not be a cheering prospect; but we shall at least be freed from the vain search for the undiscovered and undiscoverable essence of the term species.

The other and more general departments of natural history will rise greatly in interest. The terms used by naturalists of affinity, relationship, community of type, paternity, morphology, adaptive characters, rudimentary and aborted organs, &c., will cease to be metaphorical, and will have a plain signification. When we no longer look at an organic being as a savage looks at a ship, as at something wholly beyond his comprehension; when we regard every production of nature as one which has had a history; when we contemplate every complex structure and instinct as the summing up of many contrivances, each useful to the possessor, nearly in the same way as when we look at any great mechanical invention as the summing up of the labour, the experience, the reason, and even the blunders of numerous workmen; when we thus view each organic being, how far more interesting, I speak from experience, will the study of natural history become!

A grand and almost untrodden field of inquiry will be opened, on the causes and laws of variation, on correlation of growth, on the effects of use and disuse, on the direct action of external conditions, and so forth. The study of domestic productions will rise immensely in value. A new variety raised by man will be a far more important and interesting subject for study than one more species added to the infinitude of already recorded species. Our classifications will come to be, as far as they can be so made, genealogies; and will then truly give what may be called the plan of creation. The rules for classifying will no doubt become simpler when we have a definite object in view. We possess no pedigrees or armorial bearings; and we have to discover and trace the many diverging lines of descent in our natural genealogies, by characters of any kind which have long been inherited. Rudimentary organs will speak infallibly with respect to the nature of long-lost structures. Species and groups of species, which are called aberrant, and which may fancifully be called living fossils, will aid us in forming a picture of the ancient forms of life. Embryology will reveal to us the structure, in some degree obscured, of the prototypes of each great class.

When we can feel assured that all the individuals of the same species, and all the closely allied species of most genera, have within a not very remote period descended from one parent, and have migrated from some one birthplace; and when we better know the many means of migration, then, by the light which geology now throws, and will continue to throw, on former changes of climate and of the level of the land, we shall surely be enabled to trace in an admirable manner the former migrations of the inhabitants of the whole world. Even at present, by comparing the differences of the inhabitants of the sea on the opposite sides of a continent, and the nature of the various inhabitants of that continent in relation to their apparent means of immigration, some light can be thrown on ancient geography.

The noble science of Geology loses glory from the extreme imperfection of the record. The crust of the earth with its embedded remains must not be looked at as a well-filled museum, but as a poor collection made at hazard and at rare intervals. The accumulation of each great fossiliferous formation will be recognised as having depended on an unusual concurrence of circumstances, and the blank intervals between the successive stages as having been of vast duration. But we shall be able to gauge with some security the duration of these intervals by a comparison of the preceding and succeeding organic forms. We must be cautious in attempting to correlate as strictly contemporaneous two formations, which include few identical species, by the general succession of their forms of life. As species are produced and exterminated by slowly acting and still existing causes, and not by miraculous acts of creation and by catastrophes; and as the most important of all causes of organic change is one which is almost independent of altered and perhaps suddenly altered physical conditions, namely, the mutual relation of organism to organism,—the improvement of one being entailing the improvement or the extermination of others; it follows, that the amount of organic change in the fossils of consecutive formations probably serves as a fair measure of the lapse of actual time. A number of species, however, keeping in a body might remain for a long period unchanged, whilst within this same period, several of these species, by migrating into new countries and coming into competition with foreign associates, might become modified; so that we must not overrate the accuracy of organic change as a measure of time. During early periods of the earth's history, when the forms of life were probably fewer and simpler, the rate of change was probably slower; and at the first dawn of life, when very few forms of the simplest structure existed, the rate of change may have been slow in an extreme degree. The whole history of the world, as at present known, although of a length quite incomprehensible by us, will hereafter be recognised as a mere fragment of time, compared with the ages which have elapsed since the first creature, the progenitor of innumerable extinct and living descendants, was created.

In the distant future I see open fields for far more important researches. Psychology will be based on a new foundation, that of the necessary acquirement of each mental power and capacity by gradation. Light will be thrown on the origin of man and his history.

Authors of the highest eminence seem to be fully satisfied with the view that each species has been independently created. To my mind it accords better with what we know of the laws impressed on matter by the Creator, that the production and extinction of the past and present inhabitants of the world should have been due to secondary causes, like those determining the birth and death of the individual. When I view all beings not as special creations, but as the lineal descendants of some few beings which lived long before the first bed of the Silurian system was deposited, they seem to me to become ennobled. Judging from the past, we may safely infer that not one living species will transmit its unaltered likeness to a distant futurity. And of the species now living very few will transmit progeny of any kind to a far distant futurity; for the manner in which all organic beings are grouped, shows that the greater number of species of each genus, and all the species of many genera, have left no descendants, but have become utterly extinct. We can so far take a prophetic glance into futurity as to foretel that it will be the common and widely-spread species, belonging to the larger and dominant groups, which will ultimately prevail and procreate new and dominant species. As all the living forms of life are the lineal descendants of those which lived long before the Silurian epoch, we may feel certain that the ordinary succession by generation has never once been broken, and that no cataclysm has desolated the whole world. Hence we may look with some confidence to a secure future of equally inappreciable length. And as natural selection works solely by and for the good of each being, all corporeal and mental endowments will tend to progress towards perfection.

It is interesting to contemplate an entangled bank, clothed with many plants of many kinds, with birds singing on the bushes, with various insects flitting about, and with worms crawling through the damp earth, and to reflect that these elaborately constructed forms, so different from each other, and dependent on each other in so complex a manner, have all been produced by laws acting around us. These laws, taken in the largest sense, being Growth with Reproduction; Inheritance which is almost implied by reproduction; Variability from the indirect and direct action of the external conditions of life, and from use and disuse; a Ratio of Increase so high as to lead to a Struggle for Life, and as a consequence to Natural Selection, entailing Divergence of Character and the Extinction of less-improved forms. Thus, from the war of nature, from famine and death, the most exalted object which we are capable of conceiving, namely, the production of the higher animals, directly follows. There is grandeur in this view of life, with its several powers, having been originally breathed into a few forms or into one; and that, whilst this planet has gone cycling on according to the fixed law of gravity, from so simple a beginning endless forms most beautiful and most wonderful have been, and are being, evolved.

25

ATHENAEUM

Report on the 1860 Meeting of the British Association for the Advancement of Science

1860

The British Association for the Advancement of Science (BAAS) was founded in 1831 to provide a broad national audience for science. By intention, the founders of the association wanted to include people from the aristocracy as well as from the middle classes. The founders also wished to preserve harmony between religion and science. Among the active membership were clergymen from the Church of England, many drawn from the universities. While as a matter of custom women did not present papers at the meetings of the association, they were allowed to attend sessions.

In June 1860 the annual meeting of the BAAS was held at Oxford, home to one of the oldest and most distinguished universities in Europe. Darwin's Origin *had been published the previous November, so this was the first occasion on which the combined company of British scientists had the opportunity to debate in person the issues raised by the book. Darwin was not present at the meeting, though his cause was favored by the fact that his college mentor, John Stevens Henslow, was president of Section D devoted to zoology and botany. The issue of Darwin's theory came up several times during the week's events, on the most well-known occasion pitting the Bishop of Oxford, Samuel Wilberforce, against comparative anatomist Thomas Henry Huxley. Samuel Wilberforce was from a distinguished family with interests overlapping those of the Darwins and Wedgwoods; his father, William Wilberforce, had been one of the leaders in the campaign to end the slave trade. While he had made his career primarily in the church, Samuel Wilberforce had long-standing interests in natural history. Further, he had been coached for the debate by Richard Owen, the comparative anatomist. Huxley was a longtime intellectual opponent of Owen's.*

From an unsigned report from the *Athenaeum Journal of Literature, Science and the Fine Arts*, no. 1706 [July 7] (July–December 1860): 18–19.

The week which began with the Prince's[1] speech, and which has closed, under the auspices of Lord Wrottesley, with the nomination of Mr. Fairbairn, has been eminently useful, various and agreeable. Since Friday, the air has been soft, the sky sunny. A sense of sudden summer has been felt in the meadows of Christ Church[2] and in the gardens of St. John's; many a dreamer of dreams, tempted by the summer warmth, has followed the Cadiz proverb, and stealing from section A or B, has consulted his ease and taken a boat. To say that the meeting has been held in Oxford, is to say that it has been held in the midst of objects of the highest human interest and of the most delightful associations—in a city of students and professors—within reach of libraries, museums, philosophical instruments, observatories, collections of natural history such as no other provincial city in England,—or in Europe,—can boast. The hospitality has been limitless. The colleges, the private houses, have been full. The splendid and piquant New Museum has been open day and night. An unusual flutter of silk and muslin has warmed with a brighter glow the old caves of the Bodleian. Groups that Watteau would have loved to paint have been daily seen under the elms of the Broad Walk or in the shades of Magdalen. Exeter chapel, which Mr. Scott has transformed into the likeness of the Sainte Chapelle in Paris, has had its hosts of pilgrims. Every morning has brought its charming breakfast parties, every evening its charming early dinners, closed by its no less charming receptions. A splendid lecture has been given by Prof. Walker on the present state of our knowledge of the Sun; two admirable sermons have been preached at St. Mary's by Mr. Temple and Mr. Mansell, on the Religious Aspects of Science; and on Saturday night, when there was no reception at the New Museum, Dr. Daubeny received a select portion of the *savans* of both sexes in his tent at the Botanic Gardens. A batch of new Doctors of Civil Law has been added to the illustrious roll, amongst whom Prof. Sedgwick was the unquestionable lion of the day. Talking of lions reminds us that the Red Lions

[1] A number of people are mentioned in this selection. Full biographical information on them is available in the *Oxford Dictionary of National Biography* as well as on the Internet. Briefly, listed in order of mention, they are: *Prince Albert*, the prince consort, husband of Queen Victoria; *John Wrottesley*, landowner and astronomer; *William Fairbairn*, mechanical engineer and president of the BAAS in 1861; *George Gilbert Scott*, architect; *Robert Walker*, physicist and clergyman; *Frederick Temple*, educator and clergyman; *Henry Longueville Mansel* [or Mansell], theologian; *Charles Daubeny*, chemist, geologist, and botanist; *Adam Sedgwick*, geologist and clergyman; *John Spencer-Churchill*, 7th Duke of Marlborough, statesman; *George Scharf*, artist; *William Whewell*, philosopher and clergyman; *Lord Talbot de Malahide* [James Talbot], politician and archaeologist; *John Crawfurd*, orientalist and ethnologist; and *Benjamin Brodie*, chemist.

[2] References to sites at the University of Oxford occur in this excerpt. Christ Church, St. John's, and Magdalen are Oxford colleges. The Bodleian is a university library.

have had their annual feed; this time under the presidency of Prof. Huxley.[3] There have been excursions numberless; the students of Geology riding chiefly to Shotover; the lovers of Art chiefly to Blenheim. The Duke of Marlborough has paid the members of the British Association the delicate compliment of throwing open his noble grounds and galleries at the hours most convenient for their visits, and in cases where the proper applications have been made, of allowing the treasures of his private apartments to be inspected in the most liberal manner. Hundreds have accepted His Grace's generous invitation to Blenheim, where the grounds are in perfect beauty, and the glorious Raffaelles, Rubens', and Van Dycks have recently been arranged and noted by the accomplished hand of Mr. Scharf.

Yet the main interest of the week has unquestionably centred in the Sections, where the intellectual activities have sometimes breathed over the courtesies of life like a sou'-wester, cresting the waves of conversation with white and brilliant foam. The flash, and play, and collisions in these Sections have been as interesting and amusing to the audiences as the Battle at Farnborough or the Volunteer Review to the general British public. The Bishop of Oxford has been famous in these intellectual contests, but Dr. Whewell, Lord Talbot de Malahide, Prof. Sedgwick, Mr. Crawford, and Prof. Huxley have each found foemen worthy of their steel, and made their charges and countercharges very much to their own satisfaction and the delight of their respective friends. The chief cause of contention has been the new theory of the Development of Species by Natural Selection—a theory open—like the Zoological Gardens (from a particular cage in which it draws so many laughable illustrations)[4]—to a good deal of personal quizzing, without, however, seriously crippling the usefulness of the physiological investigations on which it rests. The Bishop of Oxford came out strongly against a theory which holds it possible that man may be descended from an ape,—in which protest he is sustained by Prof. Owen, Sir Benjamin Brodie, Dr. Daubeny, and the most eminent naturalists assembled at Oxford. But others—conspicuous among these, Prof. Huxley—have expressed their willingness to accept, for themselves, as well as for their friends and enemies, all actual truths, even the last humiliating truth of a pedigree not registered in the Herald's College.[5] The dispute has at least made Oxford commonly lively during the week.

[3] Huxley's "Red Lions" were an informal group of friends who met at a London pub of that name.

[4] The reference to the Zoological Gardens in London pertains to higher apes housed there.

[5] The Heralds' College is a genealogical institution.

26

ASA GRAY

Review of Darwin's Origin

1860

From the length and tone of his review, it is clear that American botanist Asa Gray (1810–1888) was fully aware of the importance of his review of Darwin's Origin. *Gray was writing for a leading American scientific journal. He was already known to many as a friendly colleague to the author being reviewed, but though generally a supporter of the theory, he had some reservations about it, particularly on philosophical and religious grounds. In the excerpts reproduced here, Gray is voicing his reservations concerning Darwin's understanding of nature. Gray hoped to align Darwin's theory with his own theistic view of nature. As Gray was in the process of writing his review, a second edition of the* Origin *appeared with what Gray termed "an additional motto" (an epigraph) on the reverse of the title page. Gray quoted the motto, and it appears at the close of the following selection.*

How the author of this book harmonizes his scientific theory with his philosophy and theology, he has not informed us. Paley [William Paley; see Document 8.], in his celebrated analogy with the watch, insists that if the time-piece were so constructed as to produce other similar watches, after the manner of generation in animals, the argument from design would be all the stronger. What is to hinder Mr. Darwin from giving Paley's argument a further *a-fortiori* extension to the supposed case of a watch which sometimes produces better watches, and contrivances adapted to successive conditions, and so at length turns out a chronometer, a town-clock, or a series of organisms of the same type? From certain incidental expressions at the close of the volume, taken in connection with the motto adopted from Whewell, we judge it probable that our author regards the whole system of nature as one which had received at its first formation the impress of the will of its Author, foreseeing the

---

From Asa Gray, "Review of Darwin's Theory on the Origin of Species by Means of Natural Selection," *American Journal of Science and Arts* 29 (1860): 182–84.

varied yet necessary laws of its action throughout the whole of its existence, ordaining when and how each particular of the stupendous plan should be realized in effect, and—with Him to whom to will is to do—in ordaining doing it. . . .

We wished under the light of such views, to examine more critically the doctrine of this book, especially of some questionable parts;—for instance, its explanation of the natural development of organs, and its implication of a "necessary acquirement of mental power" in the ascending scale of gradation. But there is room only for the general declaration that we cannot think the Cosmos a series which began with chaos and ends with mind, or of which mind is a result: that if by the successive origination of species and organs through natural agencies, the author means a series of events which succeed each other irrespective of a continued directing intelligence,—events which mind does not order and shape to destined ends,—then he has not established that doctrine, nor advanced towards its establishment, but has accumulated improbabilities beyond all belief. Take the formation and the origination of the successive degrees of complexity of eyes as a specimen. The treatment of this subject . . . , upon one interpretation is open to all the objections referred to; but if, on the other hand, we may rightly compare the eye "to a telescope, perfected by the long continued efforts of the highest human intellects," we could carry out the analogy, and draw satisfactory illustrations and inferences from it. The essential, the directly intellectual thing is the making of the improvements in the telescope or the steam-engine. Whether the successive improvements, being small at each step, and consistent with the general type of the instrument, are applied to some of the individual machines, or entire new machines are constructed for each, is a minor matter. Though if machines could engender, the adaptive method would be most economical; and economy is said to be a paramount law in nature. The origination of the improvements, and the successive adaptations to meet new conditions or subserve other ends, are what answer to the supernatural, and therefore remain inexplicable. As to bringing them into use, though wisdom foresees the result, the circumstances and the natural competition will take care of that, in the long run. The old ones will go out of use fast enough, except where an old and simple machine remains still best adapted to a particular purpose or condition—as, for instance, the old Newcomen engine for pumping out coal-pits. If there's a Divinity that shapes these ends, the whole is intelligible and reasonable; otherwise, not. . . .

The work is a scientific one, rigidly restricted to its direct object; and by its science it must stand or fall. Its aim is, probably not to deny creative intervention in nature,—for the admission of the independent origination of certain types does away with all antecedent improbability of as much intervention as may be required,—but to maintain that Natural Selection in explaining the facts, explains also many classes of facts which thousand-fold repeated independent acts of creation do not explain, but leave more mysterious than ever. How far the author has succeeded, the scientific world will in due time be able to pronounce.

As these sheets are passing through the press a copy of the second edition has reached us. We notice with pleasure the insertion of an additional motto on the reverse of the title-page, directly claiming the theistic view which we have vindicated for the doctrine. Indeed these pertinent words of the eminently wise Bishop Butler, comprise, in their simplest expression, the whole substance of our latter pages:—

"The only distinct meaning of the word 'natural' is *stated, fixed,* or *settled;* since what is natural as much requires and presupposes an intelligent mind to render it so, i. e., to effect it continually or at stated times, as what is supernatural or miraculous does to effect it for once." A. G.

27

LOUIS AGASSIZ

Review of Darwin's Origin

1860

Born in Europe, Louis Agassiz (1807–1873) became the single most prominent natural historian in the United States when he joined the faculty of Harvard University. He had been a student of Georges Cuvier and followed Cuvier in opposing evolution. Agassiz's own areas of specialization were fossil fish and glacial studies. The following excerpts are taken from his July 1860 review in the American Journal of Science and Arts, *the same journal in which Asa Gray, Darwin's most prominent proponent, had published his review earlier in the year. Agassiz's chief line of attack against Darwin pertained to the geological record. Agassiz*

From Louis Agassiz, "Comments on Darwin's *Origin* from a Forthcoming Volume of *Contributions to the Natural History of the United States," American Journal of Science and Arts* 30 (1860): 144–45, 154.

did not believe it to be as imperfect as Darwin claimed, nor did he believe that the record showed any examples of the gradual change of species over time.

Had Mr. Darwin or his followers furnished a single fact to show that individuals change, in the course of time, in such a manner as to produce at last species different from those known before, the state of the case might be different. But it stands recorded now as before, that the animals known to the ancients are still in existence, exhibiting to this day the characters they exhibited of old. The geological record, even with all its imperfections, exaggerated to distortion, tells now, what it has told from the beginning, that the supposed intermediate forms, between the species of different geological periods are imaginary beings, called up merely in support of a fanciful theory. The origin of all the diversity among living beings remains a mystery as totally unexplained as if the book of Mr. Darwin had never been written, for no theory unsupported by fact, however plausible it may appear, can be admitted in science.

It seems generally admitted that the work of Darwin is particularly remarkable for the fairness with which he presents the facts adverse to his views. It may be so; but I confess that it has made a very different impression upon me. I have been more forcibly struck by his inability to perceive when the facts are fatal to his argument, than by anything else in the whole work. His chapter on the Geological Record, in particular, appears to me, from beginning to end, as a series of illogical deductions and misrepresentations of the modern results of Geology and Palæontology. I do not intend to argue here, one by one, the questions he has discussed. Such arguments end too often in special pleading, and any one familiar with the subject may readily perceive where the truth lies by confronting his assertions with the geological record itself. But since the question at issue is chiefly to be settled by palæontological evidence, and I have devoted the greater part of my life to the special study of the fossils, I wish to record my protest against his mode of treating this part of the subject. Not only does Darwin never perceive when the facts are fatal to his views, but when he has succeeded by an ingenious circumlocution in overleaping the facts, he would have us believe that he has lessened their importance or changed their meaning. He would thus have us believe that there have been periods during which all that had taken place during other periods was destroyed, and this solely to explain the absence of intermediate forms between the fossils found in successive deposits, for the origin of which he looks to those missing links; whilst every recent progress in Geology shows more and more fully how gradual and successive all the deposits have been which form the crust of our earth.—He would have us believe that entire faunæ have disappeared before those were preserved, the remains of which are found in the lowest fossiliferous strata; when we find everywhere non-fossiliferous strata below those that contain the oldest fossils now known. It is true, he explains their absence by the supposition that they were too delicate to be preserved; but any animals from which Crinoids, Brachiopods, Cephalopods, and Trilobites could arise, must have been sufficiently similar to them to have left, at least, traces of their presence in the lowest non-fossiliferous rocks, had they ever existed at all.—He would have us believe that the oldest organisms that existed were simple cells, or something like the lowest living beings now in existence; when such highly organized animals as Trilobites and Orthoceratites are among the oldest known.—He would have us believe that these lowest first-born became extinct in consequence of the gradual advantage some of their more favored descendants gained over the majority of their predecessors; when there exist now, and have existed at all periods in past history, as large a proportion of more simply organized beings, as of more favored types, and when such types as Lingula were among the lowest Silurian fossils, and are alive at the present day.—He would have us believe that each new species originated in consequence of some slight change in those that preceded; when every geological formation teems with types that did not exist before.—He would have us believe that animals and plants became gradually more and more numerous; when most species appear in myriads of individuals, in the first bed in which they are found. He would have us believe that animals disappear gradually; when they are as common in the uppermost bed in which they occur as in the lowest, or any intermediate bed. Species appear suddenly and disappear suddenly in successive strata. That is the fact proclaimed by Palæontology; they neither increase successively in number, nor do they gradually dwindle down; none of the fossil remains thus far observed show signs of a gradual improvement or of a slow decay. . . .

. . . Were the transmutation theory true, the geological record should exhibit an uninterrupted succession of types blending gradually into one another. The fact is that throughout all geological times each period is characterized by definite specific types, belonging to definite genera, and these to definite families, referable to definite orders, constituting definite classes and definite branches, built upon definite plans. Until the facts of Nature are shown to have been mistaken by those who have collected them, and that they have a different meaning from that now generally assigned to them, I shall therefore consider the transmutation theory as a scientific mistake, untrue in its facts, unscientific in its method, and mischievous in its tendency.

Cambridge, June 80, 1860.

28

Grave Sites of Asa Gray and Louis Agassiz

Mount Auburn Cemetery in Cambridge, Massachusetts, was the first garden cemetery in the United States. Its careful landscaping was designed to create a tranquil setting for reflection. The cemetery has been designated a National Historic Landmark by the Department of the Interior. Louis Agassiz and Asa Gray are buried there. Given their common interests in natural history, it was an ideal resting place. Even in death, however, their philosophical differences are apparent. Gray, the evolutionist, was represented in death by a Christian cross; Agassiz, the opponent of evolution, was represented by a glacial erratic, recalling his service as founder of the theory of ice ages. (Erratics are boulders that have been dragged from their point of origin by the movement of glaciers.) The fact that the grave of Darwin's advocate bears the more traditional symbol suggests that in matters of science and religion, alliances were not always predictable.

Photographs by Sandra Herbert.

JEAN LOUIS RODOLPHE
AGASSIZ
ELIZABETH CARY AGASSIZ
GEORGINA SHELTON
CARY
Daughter of
RICHARD CARY
and

29

CHARLES DARWIN AND ASA GRAY

Letters

1861–1866

After the publication of On the Origin of Species, *Charles Darwin continued his correspondence with Asa Gray. Brief excerpts from their correspondence appear in this selection. In their letters the men exchanged botanical information. Gray had interested Darwin in the subject of climbing plants, and that topic was frequently discussed. They also corresponded on the state of opinion in the scientific world regarding Darwin's theory of evolution through natural selection. To further the acceptance of his theory among religious readers, Darwin arranged for the republication of some of Gray's comments advocating a theistic view of evolution. For his part, Gray kept Darwin up-to-date on his relations with Louis Agassiz, Darwin's opponent. Despite their common interest in evolution, however, the most spirited comments in the letters pertain to the American Civil War and the prospects it presented for the abolition of slavery in the United States. With the events of April 1865—the Confederate capital of Richmond fell on April 2–3, General Robert E. Lee surrendered the Army of Northern Virginia on April 9, and Abraham Lincoln was shot on April 14—came a flurry of correspondence between Darwin and Gray. Their correspondence illustrates how scientific and political interests could be intertwined. Their friendship remained strong throughout their lives and contributed to the development of evolutionary biology.*

[Charles Darwin to Asa Gray, June 5, 1861]

My dear Gray

I have been rather extra busy, so have been slack in answering your note of May 6th. . . .

From Frederick Burkhardt et al., eds., *The Correspondence of Charles Darwin*, 16+ vols. (Cambridge: Cambridge University Press, 1985–), 9:162–63; 12:47–48; 13:27, 125–26, 144–45, 207–8; 14:302, 377.

I have heard nothing from Trübner of sale of your Essay;[1] hence fear it has not been great: I wrote to say you could supply more.—I sent a copy to Sir J. Herschel; & in his new Edit of his Physical Geography he has note on the origin of species, & agrees to certain limited extent; but puts in a caution on design, so much like yours that I suspect it is borrowed.—I have been led to think more on this subject of late, & grieve to say that I come to differ more from you. It is not that designed variation makes, as it seems to me, my Deity "Natural Selection" superfluous; but rather from studying lately domestic variations & seeing what an enormous field of undesigned variability there is ready for natural selection to appropriate for any purpose useful to each creature....

But I suppose you are all too overwhelmed with public affairs to care for science.—I never knew the newspapers so profoundly interesting. N. America does not do England justice: I have not seen or heard of a soul who is not with the North. Some few, & I am one, even wish to God, though at the loss of millions of lives, that the North would proclaim a crusade against Slavery. In the long run, a million horrid deaths would be amply repaid in the cause of humanity.—What wonderful times we live in.—Massachusetts seems to show noble enthusiasm. Great God how I shd like to see that greatest curse on Earth Slavery abolished....

Farewell | Ever yours | C. Darwin

[Asa Gray to Charles Darwin, February 16, 1864]

My dear Darwin

Here we are past midwinter, and, not being stimulated as of old by your exciting letters, I have not written you a line since Christmas. Not that I have had anything in particular to tell you. I write now to say how *very sorry* I am that the word or two I get about you from [Joseph] Hooker, gives me the idea that you are having a uncomfortable and suffering time, as well as entirely broken off from scientific work....

The sentiment of our country, you must see—at least I assure you—has settled—as I knew it would if the rebellion was obstinate enough—into a determination to do away with Slavery. Homely, honest, ungainly Lincoln is the *representative man* of the country.

A Boston gentleman, at cost of $11,000 or more, is to build a fireproof house for my herbarium—which I give to the University, with

[1] Darwin had arranged a reprint of several articles by Gray from the American journal *Atlantic Monthly*.

my botanical Library. A fund of $12000 is raising to support it—which will relieve me of the expenditure of about $500 a year. But, I shall have double care and bother all the coming spring and summer. . . .

Pray, get well, dear Darwin, and believe me to be ever Yours cordially | Asa Gray

[Asa Gray to Charles Darwin, January 17, 1865]

Dear Darwin

Yours of 26th Dec. just received—long *en route*—must have crossed one from me,—yet I am not sure. . . .

People have much & many things to give for now. At present we are feeding Savannah—while the rebels are starving our men (prisoners) in the interior of the country.

Do you not begin to believe that we shall put the rebellion, restore the Union, and do away with Slavery?

Heartily do I wish you a prosperous year, and continually improving health—& power to work—and less discomfort—Also—tho' a small matter—I give you joy over the *Copley Medal*, which R.S. honors itself in giving to you.

Ever | A. Gray

[Charles Darwin to Asa Gray, April 19, 1865]

My dear Gray

I have not written for a long time (& a good job too perhaps you will think), not since receiving your letter written on Jany 19th; but I have often thought of you & often wished to write, but either had other things to do, or felt too tired. I have nothing particular to say now, but the grand news of Richmond has stirred me up to write. I congratulate you, & I can do this honestly, as my reason has always urged & ordered me to be a hearty good wisher for the north, though I could not do so enthusiastically, as I felt we were so hated by you.—

Well I suppose we shall all be proved utterly wrong who thought that you could not entirely subdue the South. One thing I have always thought that the destruction of Slavery would be well worth a dozen years war. Two days ago a very charming man, enthusiastic for the north, called here, Mr. Laugel, & he does not believe that you will attack us & Canada. I fear it will take many years before your country will shake down to its old routine.—. . .

. . . I have begun correcting proofs of my paper on "Climbing Plants." I suppose I shall be able to send you a copy in 4 or 5 weeks. I think it contains a good deal new & some curious points, but it is so fearfully long, that no one will ever read it. If, however, you do not *skim* through it, you will be an unnatural parent, for it is your child[.]

Believe me, my dear Gray, | Yours affectionately | Charles Darwin

[Asa Gray to Charles Darwin, May 15 and 17, 1865]

My Dear Darwin

Your kind letter of the 19th ult. crossed a brief note from me. I am too much *distracted* with work at this season to write letters on our affairs, and if I once begin, I should not know where to stop. You have always been sympathising and just, and I appreciate your hearty congratulations on the success of our just endeavors. You have since had much more to rejoice over, as well as to sorrow with us. But the noble manner in which our country has borne itself should give you real satisfaction. We appreciate too the good feeling of England in its hearty grief at the murder of Lincoln.

Don't talk about our "hating" you,—nor suppose that we want to rob you of *Canada*—for which nobody cares.

We think we have been ill-used by you, when you thought us weak and broken.—& when we expected better things. We have learned that we *must be strong* to live in peace & comfort with England,—otherwise we should have to eat much dirt. But now that we are on our feet again, all will go well, and *hatred* will disappear. Indeed, I see little of that. We do not even hate the Rebels, and may not even execute so much of justice as to convict of treason & hang their President, whom we have just caught,—but *I hope we shall*,—hang the leader & spare the subordinates. We are now feeding the south, who starved our men taken prisoners.

Slavery is thoroughly dead. We have a deal to do, but shall do it, I trust, and deserve your continued approbation. We have a load to carry—heavy, no doubt, but a young & re-invigorated country, with a future before it can do and bear, & prosper under what might stagger a full-grown, mature country of the Old world[.]

I must look to the *Plantago* dimorphism: for, as you say, these plants, fertilised by wind, could gain nothing by being dimorphic. No dimorphic species grows very near here,—nor can I now get seeds of P. Virginica. Perhaps a good look at even dried specimens, under your hints, may settle the matter.

I was exceedingly interested with the Lythrum paper (but had no time to write a notice of it.), & I wait expectingly for your Climbing plants. You are the very prince of investigators.

We hope presently to make Mrs. Wedgwood's[1] acquaintance.

In great haste, dear Darwin, | Your affectionate | A. Gray

[Asa Gray to Charles Darwin, July 24, 1865]

My Dear Darwin

I had heard, thro' Hooker, that you had been poorly again, and I think that a letter, signifying my sorrow was written and has crossed yours just received. I fancy you as now getting much better again. . . .

Jefferson Davis richly deserves to be hung. We are all willing to leave the case in the hands of the Government, who must take the responsibility. If I were responsible, I would have him tried for *treason*—the worst of crimes in a republic—convicted, sentenced to death,—and then I think I should commute the penalty, not out of any consideration for him, but from policy, and for his more complete humiliation. The only letters I have received expressing a desire to hang him, are from rebeldom itself—from Alabama. You see slavery is *dead,* **dead**,—an absolute unanimity as to this. The Revolted States will behave as badly as they can, but they are so thoroughly whipped that can't stir, hand or foot,—and we are disbanding all our armies—a corporal's guard is enough to hold South Carolina—

Seriously, there are difficult questions before us,—but only one result is possible—the South must be renovated, and Yankeefied.

Well—take good care of yourself, and let me know that you are again in comfortable condition[.]

Ever Your affectionate friend | A. Gray

[Asa Gray to Charles Darwin, August 27, 1866]

My Dear Darwin.

. . . Agassiz is back (I have not seen him), and he went at once down to meeting of National Academy of Sciences—from which I sedulously keep away—and, I hear *proved* to them that the glacial period covered the *whole continent of America with unbroken ice,* and closed with a

[1] *Mrs. Wedgwood*: Frances Wedgwood (d. 1874), wife of Francis Wedgwood, the owner of the Wedgwood potteries.

significant gesture and the remark "So *here is the end of the Darwin* theory"! How do you like that.

I said last winter, that Agassiz was bent upon covering the whole continent with ice,—and that the *motive* of the discovery he was sure to make was, *to make it sure* that there should be no coming down of any terrestial life from tertiary or post tertiary period to ours.

You cannot deny that he has done his work *effectually,* in a truly imperial way!

I am glad your new ed. is not to be issued for 3 months yet. I want to read the sheets at odd moments and give a notice of the new ed. in some periodical—tho' I can give little time to it.

Ever dear Darwin, | Yours cordially | A. Gray

[Asa Gray to Charles Darwin, November 6, 1866]

Dear Darwin

Thanks for yours of 19th Oct. . . .

I want you to know that Mr. Agassiz—having sent to me, thro, a mutual friend—a handsome apology for some very bad conduct to me——a mere outbreak of spoiled temper, 1½ years ago, we are now on amicable terms—Till then, I was obliged to ignore him.

Ever Yours | A. Gray

A Chronology of the History of Evolutionary Ideas (1739–1882)

1739 Georges-Louis Leclerc, Comte de Buffon, appointed director of the Jardin du Roi/Muséum d'Histoire Naturelle, a post he holds until his death in 1788.

1787 Erasmus Darwin's *Families of Plants*, his translation of Carl Linnaeus's *Genera Plantarum*, published.

Thomas Jefferson's *Notes on the State of Virginia* published in London.

1790s Georges Cuvier establishes species extinction as a fact from his study of fossil vertebrates.

1798 Thomas Robert Malthus's *An Essay on the Principle of Population* published.

1799–1804 Alexander von Humboldt travels to the New World.

1800 Jean-Baptiste Lamarck suggests in lecture that species may have altered rather than become extinct.

1802 William Paley's *Natural Theology* published.

1803 Erasmus Darwin's *Temple of Nature* published posthumously.

1807 Geological Society of London established.

1809 Lamarck's *Philosophie zoologique* published.

Charles Darwin born.

1812 Cuvier's four-volume work on fossil quadrupeds published.

1830–1833 Charles Lyell's *Principles of Geology* published.

1831–1836 Darwin serves as naturalist aboard the surveying ship HMS *Beagle*.

1837 Darwin privately adopts the theory of transmutation.

1838 Darwin reads Malthus's *An Essay on the Principle of Population.*

1839 Darwin's *Journal of Researches*, about his *Beagle* voyage experiences, published.

1842, 1844 Darwin privately writes drafts of his theory of evolution through natural selection.

1844 Robert Chambers's *Vestiges of the Natural History of Creation* published anonymously.

1845 Darwin's second edition of his *Journal of Researches* published.

1858 Alfred Russel Wallace sends Charles Darwin a manuscript describing his own theory of evolution; Wallace's essay on evolution published in the *Journal of the Linnean Society of London*, together with some of Darwin's early writings on evolution.

1859 Darwin's *On the Origin of Species* published.

1860 Debate begins in Britain, the United States, and elsewhere over Darwin's theory of evolution through natural selection.

1864 Darwin awarded the Copley Medal by the Royal Society of London.

1868 Darwin's *The Variation of Animals and Plants under Domestication* published.

1870–1871 Darwin's *The Descent of Man* published.

1872 Darwin's *The Expression of the Emotions in Man and Animals* published.

1882 Darwin dies.

Questions for Consideration

1. What was Carl Linnaeus's major contribution to the study of natural history?
2. How did foreign travel help European researchers understand that species might be mutable?
3. What role did a belief in species extinction play in the debate over evolution?
4. How did the Muséum d'Histoire Naturelle enhance Georges Cuvier's and Jean-Baptiste Lamarck's understanding of species?
5. What sort of transatlantic ties were apparent within the natural history community in the eighteenth century?
6. Erasmus Darwin's poetic vision joined nature and politics, as reflected in the title of his book *The Temple of Nature; or, the Origin of Society*. From his point of view, what did nature and society have in common?
7. Two thinkers writing outside the natural history tradition—Thomas Robert Malthus and William Paley—influenced Charles Darwin's understanding of species change. Which of their ideas were most influential to Darwin?
8. Compare and contrast the contributions of Georges Cuvier and Jean-Baptiste Lamarck to natural history.
9. Please explain the following statement: Although strongly antievolutionary in his views, geologist Charles Lyell set up the problem of species in a way that it could be solved.
10. How would you rate Darwin's experiences aboard the *Beagle*: as essential to his adoption of evolutionary views, as hastening his adoption of those views, or as incidental to his views?
11. What role did the abolitionist movement play in Darwin and Wedgwood family history? Were Darwin's ideas on abolition and his ideas on the evolution of species at all related, and if so, how?
12. How were views on human nature affected by transmutationist thought?

13. What was Emma Darwin's perspective on religion? What did she think of her husband's views?
14. Why did Charles Darwin import Thomas Robert Malthus's views into his theory?
15. What role did artificial selection play in Darwin's theory of evolution through natural selection?
16. What role did geology play in the debate over evolution from the 1780s to the 1860s?
17. In what ways were Alfred Russel Wallace's and Charles Darwin's experiences in life similar or different? Did these similarities or differences manifest themselves in their published views?
18. Did Charles Darwin behave ethically in his handling of the manuscript on evolution that Wallace sent him in 1858?
19. Would Darwin have done better to center his *Origin of Species* on the study of variation in one group, such as pigeons?
20. Does the argument in *Origin of Species* make logical sense? What are the strengths and weaknesses of the argument?
21. Compare and contrast the initial reception of Darwin's *Origin* in Britain and in the United States.
22. In the contrasting reviews of the *Origin* by Asa Gray and Louis Agassiz, which man made the better case?
23. In what ways were politics and science intertwined in regard to the theory of evolution between the 1780s and the 1860s?
24. What views in politics and science did Gray and Darwin share? Were there issues on which they differed?
25. Where did Darwin think his theory of evolution through natural selection would lead?
26. Would you agree with Lyell's high estimate of Darwin's "capacity as a thinker and a philosophical writer"? Please explain.

Selected Bibliography

There are several comprehensive works useful to the student. One well-regarded source is Peter J. Bowler, *Evolution: The History of an Idea*, 3rd ed. (Berkeley: University of California Press, 2003). Ernst Mayr's *The Growth of Biological Thought: Diversity, Evolution, and Inheritance* (Cambridge, Mass.: Harvard University Press, 1982) also provides an expert and accessible survey with discussion of numerous individuals unfamiliar to most historians. Two additional well-written general works are John C. Greene, *The Death of Adam: Evolution and Its Impact on Western Thought*, rev. ed. (Ames: Iowa State University Press, 1996), and Michael Ruse, *The Darwinian Revolution: Science Red in Tooth and Claw*, 2nd ed. (Chicago: University of Chicago Press, 1999).

On the Lunar Society, including Erasmus Darwin, see Jenny Uglow, *The Lunar Men* (New York: Farrar, Straus and Giroux, 2002). On the eighteenth-century paleontological background, including a superb treatment of Georges Cuvier, an indispensable broad survey is Martin Rudwick, *Bursting the Limits of Time: The Reconstruction of Geohistory in the Age of Revolution* (Chicago: University of Chicago Press, 2005). Continuing the story through the mid-nineteenth century is the same author's *Worlds before Adam: The Reconstruction of Geohistory in the Age of Reform* (Chicago: University of Chicago Press, 2008). On the search for prehistory in Britain, see A. Bowdoin Van Riper, *Men among the Mammoths* (Chicago: University of Chicago Press, 1993). For an intriguing look at contributions of native Americans to paleontological knowledge, see Adrienne Mayor, *Fossil Legends of the First Americans* (Princeton, N.J.: Princeton University Press, 2005).

On the continental European history of evolutionary ideas, including a rich appreciation of Alexander von Humboldt, see Robert J. Richards, *The Romantic Conception of Life: Science and Philosophy in the Age of Goethe* (Chicago: University of Chicago Press, 2002). On early French-American differences with regard to natural history, see Lee Alan Dugatkin, *Mr. Jefferson and the Giant Moose* (Chicago: University of Chicago Press, 2009). Concerning the reception of evolutionary ideas in the United States, see Ronald L. Numbers, *Darwinism Comes to America* (Cambridge, Mass.: Harvard University Press, 1998). Thomas Glick has also written extensively on the comparative history of the reception of Darwinism, as, for example, his widely used *The Comparative Reception of Darwinism* (Chicago: University of Chicago Press, 1988).

For studying the contributions of individuals discussed in this book, a student with limited time in the library would do well to begin with entries in standard biographical dictionaries, including the *Oxford Dictionary of National Biography*, the *American National Biography*, and the *Dictionary of Scientific Biography*. Reference librarians are able to guide students to these sources.

Many of the figures discussed in this book have received full-length biographical treatment. On John Stevens Henslow, see S. M. Walters and E. A. Stow, *Darwin's Mentor* (Cambridge, U.K.: Cambridge University Press, 2001). On Charles Lyell, see Leonard Wilson, *Charles Lyell: The Years to 1841* (New Haven, Conn.: Yale University Press, 1972), and its sequel *Lyell in America* (Baltimore: Johns Hopkins University Press, 1998). On Robert Chambers, see James A. Secord, *Victorian Sensation* (Chicago: University of Chicago Press, 2000). On Alfred Russel Wallace, begin with Martin Fichman, *An Elusive Victorian* (Chicago: University of Chicago Press, 2004). On Joseph Hooker, see Jim Endersby, *Imperial Nature* (Chicago: University of Chicago Press, 2008). For Louis Agassiz and Asa Gray, there are two classic treatments: Edward Lurie, *Louis Agassiz* (Baltimore: Johns Hopkins University Press, 1988), and A. Hunter Dupree, *Asa Gray* (Baltimore: Johns Hopkins University Press, 1988). On Thomas Henry Huxley, see Adrian Desmond, *Huxley* (Reading, Mass.: Addison-Wesley, 1997). For a comparison of Huxley and Richard Owen, see Christopher E. Cosans, *Owen's Ape and Darwin's Bulldog* (Bloomington: Indiana University Press, 2009).

For Darwin himself, there is an authoritative up-to-date guide to the literature available: Michael T. Ghiselin, *Darwin: A Reader's Guide* (San Francisco: California Academy of Sciences, 2009). Two fine biographies were written by well-known historians of science: Adrian Desmond and James Moore, *Darwin* (New York: Warner Books, 1991) and Janet Browne, *Charles Darwin: Voyaging* (New York: Alfred A. Knopf, 1995) and its sequel *Charles Darwin: The Power of Place* (New York: Alfred A. Knopf, 2002). On the *Beagle* voyage, see Frederick Burkhardt, ed., *The 'Beagle' Letters* (Cambridge, U.K.: Cambridge University Press, 2008). For an anthropological perspective on Darwin's experience with native peoples during the voyage, see Anne Chapman, *European Encounters with the Yamana People of Cape Horn* (Cambridge, U.K.: Cambridge University Press, 2010). Also on the voyage experience, see K. Thalia Grant and Gregory B. Estes, *Darwin in Galápagos* (Princeton, N.J.: Princeton University Press, 2009). For recent scholarly work surrounding the *Origin*, see Michael Ruse and Robert J. Richards, eds., *The Cambridge Companion to the "Origin of Species"* (Cambridge, U.K.: Cambridge University Press, 2009). On the aftermath of the *Origin*, see Frederick Burkhardt, Samantha Evans, and Alison Pearn, eds., *Evolution: Selected Letters of Charles Darwin, 1860–1870* (Cambridge, U.K.:

Cambridge University Press, 2008). On the geological aspect of Darwin's work, see Sandra Herbert, *Charles Darwin: Geologist* (Ithaca, N.Y.: Cornell University Press, 2005). Regarding the issue of Darwin and slavery, see Adrian Desmond and James Moore, *Darwin's Sacred Cause* (Boston: Houghton Mifflin Harcourt, 2009). For an engaging treatment of Darwin and Lincoln in juxtaposition, see Adam Gopnik, *Angels and Ages* (New York: Alfred A. Knopf, 2009). On the role of Darwinism in post–Civil War American intellectual life, consult Louis Menand, *The Metaphysical Club* (New York: Farrar, Straus and Giroux, 2001).

There are three important Web resources for Darwin: the Darwin Digital Library of Evolution (http://darwin.amnh.org), the Darwin Correspondence Project (www.darwinproject.ac.uk), and the Complete Work of Charles Darwin Online (www.darwin-online.org.uk). On natural history, a useful Web site is the Encyclopedia of Life (www.eol.org).

Acknowledgments (continued from p. iv)

Document 11: Walter F. Cannon, "The Impact of Uniformitarianism: Two Letters from John Herschel to Charles Lyell, 1836–1837," *Proceedings of the American Philosophical Society* 105 (1961): 304–5. Reprinted with permission of the American Philosophical Society.

Document 14: Nora Barlow, ed., "Darwin's Ornithological Notes," *Bulletin of the British Museum* (Natural History) Historical Series 2, no. 7 (1963): 262. Reprinted with permission of the Natural History Museum, London.

Document 15: Paul H. Barrett, Peter J. Gautrey, Sandra Herbert, David Kohn, and Sydney Smith, eds., *Charles Darwin's Notebooks, 1836–1844: Geology, Transmutation of Species, Metaphysical Enquiries* (Ithaca, N.Y.: Cornell University Press; London: British Museum (Natural History) and Cambridge, U.K.: Cambridge University Press, 1987), 171, 177, 195. Copyright 2008 by the Committee for the Publication of the Charles Darwin's Notebooks. Reprinted with the permission of Cambridge University Press.

Document 16: Frederick Burkhardt et al., eds., *The Correspondence of Charles Darwin* (Cambridge, U.K.: Cambridge University Press, 1985–), Vol. 2: 171–73. Reprinted with the permission of Cambridge University Press.

Document 20: Frederick Burkhardt et al., eds., *The Correspondence of Charles Darwin* (Cambridge, U.K.: Cambridge University Press, 1985–), Vols. 6: 447–50 and 7: 507–11. Reprinted with the permission of Cambridge University Press.

Document 23: Frederick Burkhardt et al., eds., *The Correspondence of Charles Darwin* (Cambridge: Cambridge University Press, 1985–), Vol. 7: 288–91. Reprinted with the permission of Cambridge University Press.

Document 29: Frederick Burkhardt et al., eds., *The Correspondence of Charles Darwin* (Cambridge, U.K.: Cambridge University Press, 1985–), Vols. 9: 162–63; 12: 47–48; 13: 27, 125–26, 144–45, 207–8; 14: 302, 377. Reprinted with the permission of Cambridge University Press.

Index